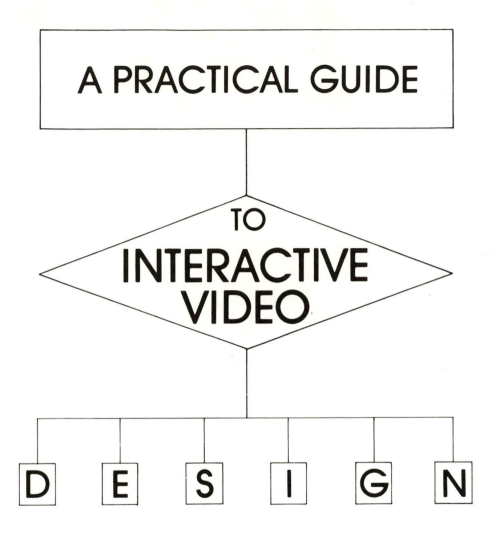

A PRACTICAL GUIDE

TO
INTERACTIVE VIDEO

D E S I G N

by Nicholas V. Iuppa

Knowledge Industry Publications, Inc.
White Plains, NY

Video Bookshelf

A Practical Guide to Interactive Video Design

Library of Congress Cataloging in Publication Data

Iuppa, Nicholas V.
    A practical guide to interactive video design.

    Bibliography: p.
    Includes index.
    1. Computer-assisted instruction.   2. Video tapes in
education.   I. Title.
LB1028.5.I84 1984   371.3'9445   84-7872
ISBN 0-86729-041-2

Printed and bound in the United States of America

Copyright © 1984 by Knowledge Industry Publications, Inc., 701
Westchester Ave., White Plains, NY 10604.

10 9 8 7 6 5 4 3 2

Contents

List of Tables and Figures

Foreword

Interactive video is an emerging technology being shaped by experts from a variety of disciplines. Not all of these experts have the same orientation or goals and, as a result, there is a good deal of controversy regarding the definition and applications of the medium.

This book is based on my experiences with interactive media during the past 14 years. It takes a rather strong position on some issues, and may add to the controversy surrounding this new technology, although that is not its intent. The book reflects my background, which has been in very pragmatic media production shaped largely by the demands of clients who expect good quality, rapid turnaround and low cost.

My involvement in interactive media began while I was working for Eastman Kodak as a designer and developer of instructional material in the late 1960s. At the time, Kodak was experimenting with individualized interactive instruction. We combined several slide carousel projectors with a big mainframe computer (there were no minis or micros at the time). Work at Kodak convinced me that interactive instruction was a powerful tool which would eventually dominate the educational process.

Later, as head of the Bank of America's instructional design unit, I supervised the development of several hundred training programs using the principles set forth by Joe Harless, the educational technologist. These principles are referred to throughout this book and form the basis for the section on instructional design.

My commitment to interactive video took shape during my last five years at Bank of America. As head of its video production facility, I had to produce video training tapes for a variety of clients from within the organization. Our clients expected tapes that would teach people how to perform detailed procedures with a very high degree of accuracy. I became more and more certain that the linear video programs we were producing just did not have the ability to do that.

1

As my frustration with linear video grew, I sought a system that would allow us to teach *skills* with video. The result of this effort was the creation of two complex experimental interactive video disc programs, which are described in this book.

Presently, as vice president of media production for ByVideo, Inc., a company that creates integrated video shopping systems, I am producing interactive video discs full time.

I have drawn upon all of these experiences to form the foundation for this book. My purpose throughout is to offer an approach to interactive video that will help it become the most powerful communications medium yet invented. To do that, I feel that the system must be driven by all the power and effectiveness of television. There is heavy emphasis in this book on the need to be entertaining, and on the proper design and implementation of large instructional packages.

It is true that experimentation with interactive video will continue. Breakthroughs in video and computer technology will affect its future growth and development. However, I believe that the programs produced today will play an important role in the medium's acceptance in the future. For that reason, I will push for high production values, creative freedom and, above all, any techniques that will make things clearer to users. I think all are essential. See if you don't agree.

Nick Iuppa
May 1984

1

What Interactive Video Is

OVERVIEW

One of my earliest childhood memories is of a radio adaptation of "The Veldt," an episode in the science fiction book *The Illustrated Man,* by Ray Bradbury.* The story involves a futuristic house which offers its well-to-do owners the ultimate baby-sitter: a playroom whose walls are floor-to-ceiling television screens. The room creates dozens of different environments for the children to play in. The children control their play environment by sending orders with their minds telepathically.

I didn't realize it at the time, of course, but the story does provide an excellent introduction to the concept of interactive video. In the story, the children were surrounded by a video environment which they selected; moreover, they controlled the sequence of events. However, they did not control the effects of their choices: the interactive system answered their directives with responses of its own.

While the concept of changing your physical environment by sending thought commands is still in the realm of science fantasy, the idea of affecting the outcome of events on a television screen is no longer fiction. Interactive video instruction has the ability to provide different outcomes depending on different user input. That capability represents an excellent way to let students, trainees and other users see the consequences of their choices.

---

*Ray Bradbury, *The Illustrated Man* (New York: Doubleday & Co., 1958).

3

Varied Outcomes

Part of what interactive video is, then, is a technique of *varied outcomes*. What is more, the outcomes have an immediate, visual representation on the TV screen. This ability to show users the consequences of their actions is what makes interactive video a powerful new teaching tool, and, in fact, much of this book will deal with interactive video's benefits for training. However, the technique can be used for very different applications as well.

It can make a dramatic difference in point-of-sale presentations, as well as in many kinds of home entertainment programs. For example, sword and sorcery and similar role-play games all take you to a point where decisions must be made— imagine taking the wrong turn and coming face-to-face with video demons rendered with all the power of television!

Literature, too, may be affected by interactivity. For years, scholars have delighted in discussing the different outcomes that may have been possible had Hamlet been a little more decisive. Future Shakespeares may need to offer not one, but many scenarios that show every possible outcome of important decisions.

Random Access

The technology that permits varied outcomes is the marriage of the computer and video. Together, these technologies offer *random access,* the ability to retrieve any piece of information easily and rapidly regardless of its location in a computer or video program.

Michael Naimark, a San Francisco cinematographer and a participant in the Massachusetts Institute of Technology's "Movie-Maps Project," likes to compare the emergence of random access interactive video to the moment when people switched from scrolls to books. Until then, people had to read scrolls in order, from beginning to end. Ceremonies evolved which were related to the unwinding of the scroll. The fact is, no one can jump to the last page of a scroll—there are no pages. Even if a person wanted to roll to the last paragraph of a scroll, he or she would have to roll past all the previous paragraphs. Books changed this. The basic functional design of a book differed from that of a scroll: people could turn to the last paragraph of a book without even seeing earlier pages. In this sense, books really were the first random access medium.

Random access, of course, is a feature of any kind of printed reference material. You don't have to go from page one to page two, or volume one to volume two. If you want to find out about music, you look in the encyclopedia's "M" book, on the page where music is listed. Many of us like to dive into the middle of a magazine to read the article we want—we don't have to begin reading from page one every time. We are accustomed to the ability to *access* any spot at random on any print piece.

The same does not hold true for linear TV. We can hunt for a scene on a video tape every now and then, but it is a time-consuming and inaccurate endeavor. The television medium as it is used today is just not designed for random access. The curious thing is that few people seem to recognize that perhaps linear presentation is limiting. However, there are signs that this situation is changing.

Steve Poe, the futurist and former head of marketing for ARDEV's video disc effort, likes to talk about a time when video software—like today's text-oriented software—will be divided equally between fiction/entertainment programming and non-fiction/reference programming. "How many old movies can we watch before we're fed up?" he asks, and he is not alone. Many people believe that the demand for non-fiction video will eventually exceed that for today's linear video entertainment.

The technology for all of this is already in place. But technology is only one part of the process; the other is proper program design. Design is what separates good interactive video programs from ineffective ones. It also distinguishes true interactive video from video-enhanced computer assisted instruction (CAI). Let's explore this distinction further.

INTERACTIVE VIDEO—A DEFINITION

Trying to define interactive video can be tricky. On the one hand, the definition must take into account interactive systems so elaborate that they would take 50 pages to explain. On the other, some systems live up to a limited definition that is simplistic enough to include nearly any two-way communications system, regardless of the quality of the interaction or the role played by video in the system.

For the purposes of this book, interactive video is defined as *any video system in which the sequence and selection of messages is determined by the user's response to the material.* The key phrase here is *video system,* a point which needs to be examined further.

Interactive Video vs. Video-Enhanced CAI

The reason that this distinction is important is that it hits you right in the pocketbook. There were many computer design professionals who built CAI systems that were not successful. By adding some video footage or sequences to what is essentially the same computer program, they are now able to call their systems interactive video. The unsuspecting executive, trying to introduce an interactive video system and to start laying the groundwork for an even greater expansion of the system, often gets a slick sales pitch from a vendor, sees a little demonstration and ends up shelling out hundreds of thousands of dollars for a computer-based instruction system with a little video on the side.

It seems to me that someone purchasing an interactive video system should get as much out of the capabilities of the system as possible. At the very least, an executive should know what interactive video's true capabilities are in order to be able to make an informed decision.

With this in mind, let's build an interactive video system that demonstrates the different roles that can be played by the computer and video and also suggests the levels of complexity you can reach with interactive video programming. The type of programming discussed here is intructional or training programming.

The simplest instructional program includes four elements: 1. a presentation of information; 2. a method by which the student can offer a response to questions about the information; 3. a mechanism for selecting a reaction to the response; and 4. the presentation of that reaction. These elements can be translated into more familiar terms: the presentation of information is known as the *demonstration;* the student's response opportunity is called an *exercise* item; the way the system selects the appropriate response to the input is called *control*; and the presentation of the system's response is called *feedback*.

These elements, and the sequence in which they occur, are normally represented in a diagram, or flowchart. A flowchart is a shorthand method used by programmers to present information in a simplified way. Flowcharts are discussed in detail throughout this book. Figure 1.1 represents a simple flowchart for interactive video instruction.

In Figure 1.1, every element that appears in a box is some kind of a display (something that the system will show, such as pages of text on a screen or segments with video motion). The control element appears in a diamond instead of a box, because it represents a decision that the computer makes rather than a display seen by the user.

In the higher forms of interactive video, part or all of the display segments are presented as video; the rest is done by the computer. The weakest interactive video programs only have a small part of the display done as video, the rest is in computer text or graphics. I prefer to think of these types of programs as video-enhanced computer assisted instruction since video plays such a small role in the program.

Notice again that in Figure 1.1 the exercise items and the feedback are displays. Now remember, at the exercise step, what a learning system does is to offer the student a way to respond, often through a series of choices. There is no reason that this step has to be presented as computer text; in fact I think that a great deal of realism is lost when it is.

To get a clear response from the student, it is important to offer a realistic simulation of whatever is being learned. It is particularly important to present exercises featuring as much realism as possible. Pages of computer text do not really "simulate"

Figure 1.1: Simple Flowchart of Instructional Program Elements

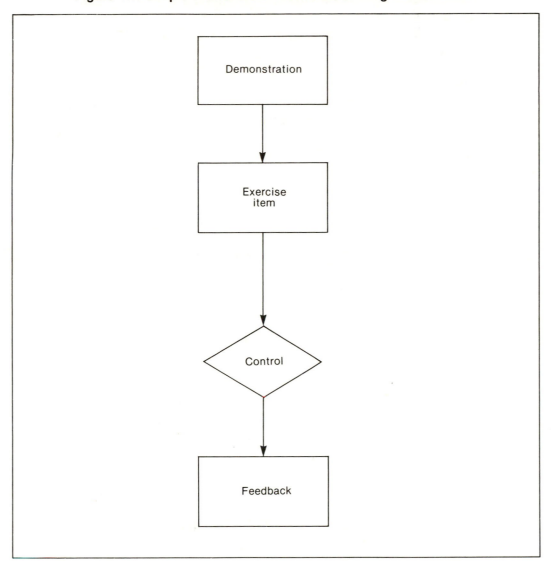

any type of activity. Therefore, good interactive video programs do not rely on the computer to present exercise items.

The same point can also be made about feedback. While it is possible to present feedback in the form of computer text on the screen, it is seldom wise to do so. The reason has to do with the essence of the interactive video experience—its *quality*. If you merely show a small video demonstration, followed by a computer-text display of multiple-choice questions, then reply to the student's response with a computer-generated text frame, you are not making the best use of the system.

The essence of the interactive video experience is video simulation, with as much video presentation of real images as possible. Whenever a supposedly interactive video system shows a page of computer text describing something that could be better shown as a video sequence, it is compromising the medium, lessening its impact and giving the customer something less than he or she should be getting.

The Role of the Computer

What is the computer's purpose in an interactive video program? Its major function is the item in the diamond in the flowchart in Figure 1.1—control. The computer is, essentially, the brains of the interactive system. Disc manufacturers and producers have divided discs into four levels according to the way the discs deal with computer control. Where the control program resides determines the complexity of the disc. Table 1.1 identifies the four disc levels.

Table 1.1: Video Disc Levels

Level	Description
I	An interactive disc with no built-in programming for computer control
II	An interactive disc with the computer program built in
III	An interactive video disc player controlled by an external computer
IV	An interactive disc that is part of a larger system of information retrieval, e.g., banks of video disc and tape players that are connected online to a central computer

Let's keep this definition of interactive video and its levels in mind as we examine interactive video equipment.

INTERACTIVE VIDEO EQUIPMENT CONFIGURATIONS

There are a variety of equipment configurations used to present interactive video programs. They range from the classic microcomputer combined with 1/2-inch video tape player systems to systems involving several video disc machines, a specially constructed outboard computer, a character generator and a printer all enclosed in a special kiosk. To help you determine which configuration is best for your uses, several options and their advantages and disadvantages are discussed below.

Laser Video Disc vs. Video Tape

Technically and mechanically, an interactive laser video disc system is superior to an interactive video tape system. The advantages of the laser disc over video tape are:[1]

---

[1]For an expanded discussion, see Steve and Beth Floyd, *Handbook of Interactive Video* (White Plains, NY: Knowledge Industry Publications, Inc., 1982).

- Rapid and precise access to any frame—searches are done much more rapidly on video disc than on tape;

- Unlimited still frame capability with far greater clarity than on tape;

- No wear and tear on the recorded medium—the disc is virtually indestructible;

- Extremely low duplicating costs for large numbers of copies—providing economies of scale that video tape cannot match;

- Greater storage density.

Video tape can become an interactive medium, but the only reason for doing so, given the above drawbacks, is economics. Video discs are expensive to master (anywhere from $1500 to $2500 or more per program); video tape dub masters cost very little. Also, the price of video tape recorders has been dropping, in some cases well below that of disc players. If you already have a personal microcomputer, a 1/2-inch solenoid-driven video tape cassette player and an interface, you can save the cost of converting to an interactive video disc system. However, you do it by sacrificing clean, efficient and, therefore, effective operation.

Laser Video Disc vs. Other Disc Systems

To date, no system except the laser video disc has proven to have the random access capability necessary for interactive video. It should be noted that there are still several laser video disc systems in various stages of development, but only one is on the market in the U.S. as of this writing: the optical reflective laser video disc, manufactured by N.V. Philips, Pioneer and Sony.

With regard to disc systems using contact technologies, only one has random access—the VHD disc introduced by JVC—and is not currently being marketed in the U.S. RCA, which was reportedly developing random access capability for its capacitance electronic disc (CED) system, withdrew from the video disc market in early 1984.

On-Board Microprocessor vs. External Computer

Several disc players feature a microprocessor built into the player to control the flow of the program, branching, responses and other interactions. The two instructional interactive video programs produced at the Bank of America used Pioneer and Sony disc players which made use of on-board microprocessors. (These kinds of programs are often referred to as level II programs.)

The biggest advantage of an on-board microprocessor is that it makes a $200 to $3000 external computer unnecessary. This can often become an important consideration. For example, a video system the size of Bank of America's (1100 players) would require 1100 disc players and 1100 outboard computers of the Apple III variety to make a system-wide conversion to interactive video. We would tolerate a good many production difficulties to save the cost of those computers—roughly $2 million dollars.

Just what kinds of production difficulties would you face if you used an on-board microprocessor-controlled system? To begin with, the program that controls the branching on the level II disc is contained in audio channel #2 and is put on the disc when it is manufactured. Because the on-board microprocessor uses a rather primitive machine language it is very often programmed by the manufacturer. The result is greater manufacturing cost, greater chance of error (because there are more people involved in the communication process) and possibly longer timetables.

Even if programmed level II discs did not have time and cost drawbacks, the on-board microprocessor system still presents several limitations in the area of updating and data collection.

The computer program in an outboard computer can be modified at any time to add new graphics, to change statistics or to skip over outdated or outmoded sequences. This is a major benefit when dealing with volatile material such as schedules, prices, interest rates, etc. With an on-board microprocessor, this information will be on the video disc and can only be changed by making a new disc. An outboard computer lets you change this data without remaking the video disc.

An outboard computer also has the ability to capture, store and analyze data that are usually lost to on-board microprocessors. A good deal of market research, instructional evaluation, product, buyer or student data can be obtained. These data will allow you to improve the program, keep track of users and generally make your effort more effective and successful.

Still Frame vs. Motion

This issue may seem to be more a matter of style than of hardware configuration, and to some extent it is. Still frame and video both have their place in interactive video programs: the trick is not to overuse still frames simply because the laser disc has such a good still frame capacity.

Video and movies are distinguished from other media by motion. Teaching people how to do anything that has to do with motion is greatly aided when you can actually show motion. Interpersonal skills, manual skills, even many cognitive skills, are best taught by motion sequences. Video sales efforts also benefit from motion. A 50,000-page still frame catalog may sound appealing, but, as many people say, you can't sell the steak without the sizzle—and in this case, motion provides the sizzle. Obviously, this is particularly important if the product you are selling involves motion —e.g., cars, champagne, video games, etc.

The important thing to keep in mind is that the choice between still frame and motion affects your hardware configuration. Therefore, don't select a configuration that severely limits your ability to offer both motion and still frame. The laser video disc provides both exceptionally well. Interactive video is a unique format that requires both motion *and* stills to provide the person-machine interaction that is the essence of the system. Don't give up either.

Computer Graphics vs. Video Graphics

This again is a question of style and, like the still frame versus motion discussion, there are both technological and economic implications.

First, it is impossible to conceive of anyone actually preferring computer lettering over video lettering. Most people, raised as they are with television, choose video lettering and titles every time. Therefore, relegate the titling job to the video system whenever you can. However, there may be some cases in which you simply must rely on computer-generated titles. These include the following:

- Whenever the information is subject to change. For example, pricing information, interest rates, procedures that might change and whole question-and-answer cycles based on volatile information.

- Whenever the system is displaying words or numbers entered by the user, such as order information in a point-of-sale system.

- Whenever a teaching exercise requires simulation of typing or word-entering behavior that cannot be simulated by multiple-choice or other limited keypad exercises.

- For games or behavior simulations that require lightning-quick response time. The trick is to lay the computer image over the video image in such a way that it appears to be part of the video image.

In all other cases rely on the clarity, flexibility and readability of video titles and graphics, unless you're on a real shoestring budget—in that case, computer-generated graphics are better than nothing.

Single Player vs. Multiple Player Systems

The advantages of multiple player systems are really quite simple and fall into a very limited range of applications. Multiple machine systems primarily enhance branching options and greatly increase the speed and flexibility of accessing information. If your goal is the total elimination of black frames, multiple machine systems let you cut between various options and offer the closest possible simulation of live-action decision making.

Another major advantage of multiple player systems is increased disc capacity, or "real estate" as it is called. You get 100,000 instead of 50,000 frames; an hour instead of (roughly) a half an hour. This added capacity is best suited to large information centers, such as the one at EPCOT Center, in Disneyworld. For more modest training and point-of-sale applications, a single laser disc player should do just fine.

Touch Screens, Joysticks and Keypads

This is not really a question of comparison and debate. Each item has its own purpose and, in many cases, one item cannot replace the other.

The installations of interactive disc systems at EPCOT Center give a strong indication that whenever a general purpose control is needed, touch screens are favored, especially for public contact uses such as information delivery. Of course, touch screens also have drawbacks. They are expensive, somewhat delicate and require more sophisticated computer programming than other controls.

Joysticks, track balls and buttons are most often used for video games. Here, these controls frequently match the objectives of the game. Small portable keypads appear to be favored over large typewriter-style keyboards for applications in which a teacher leads a class through interactive group experience.

The main point is that controls are dictated by the applications, so you must become familiar with all types of controls. Common controls, such as the computer keyboard and keypad, are fine first steps, but broader applications require more sophisticated controls.

Computer Audio vs. Disc Audio

Video disc systems produce some of the best audio in the world; computer audio is a weak synthesized version of real sounds. Therefore, it is unlikely that you would ever choose computer audio, except for accents, effects or limited applications—for example, when computer music covers a disc search, paging or reverse action sequence.

Actually, a more interesting debate is whether a disc should use any audio or none at all. Should a point-of-sale system sit there buzzing and gurgling until it is put into use, or will an interesting video display be just as attractive and far less annoying to potential customers? Also, should frames be displayed silently on the screen, or will reading them to students encourage attention and learning?

Although I have no hard statistics, I have encountered very positive feedback and statements of preference for the use of voice-overs during the display of titles or menus as still frames. Adding audio does take up disc space, but in some cases it may be worth it. On the other hand, for point-of-sale presentations, cycles set up to attract users should remain silent. There is no point in turning the point-of-sale location into a penny arcade. In addition, most malls, stores, hotels and other point-of-sale locations are likely to object.

ABOUT THIS BOOK

The rest of this book is devoted to presenting a practical instructional and production framework which readers can use when designing interactive video programs. It is aimed particularly at instructional designers, writers, producers, programmers and directors of interactive video programs. Its objectives are to:

- Explain the principles used in design and development of all levels of interactive video programming;

- Present design options to those engaged in interactive video production;

- Illustrate the benefits of interactive technology, whenever possible through examples based on practical experience;

- Inspire the designers and producers to press the limits of the medium in order to discover its most beneficial applications and take advantage of all interactive video can be.

Throughout this book, I will refer to several interactive programs that I have been involved with. They are described in Table 1.2, which appears on page 14.

As mentioned earlier, the sometimes strong points of view expressed in this book are based on specific experiences with these programs. I make no apology for the opinions and recommendations contained herein, since I feel they will ultimately serve the production community and emerging interactive production industry. There will be some who disagree with my positions, and to those I say: Let the dialogue begin.

Table 1.2: Interactive Program Examples

Program	Description	Audience	Year	Credits
Debits and Credits	A 27-minute interactive program to teach hard skills on the difference between debits and credits and the use of teller stamps.	New tellers	1981	Director: Russ Srole Writer: Nicholas V. Iuppa Programmer: Gary Giddings Disc manufacturer: DiscoVision Production house: The Malibu Group
People Skills	A 4-sided, 90-minute course on interpersonal relations including: greeting customers, listening for content and feelings, clarifying content and feelings, getting down to business. Program used a soap opera story line about a teller's marital problems, together with a wide range of customers. They ranged from typical people to "far out" humorous types. Instructional design and exercise construction was meant to demonstrate the fullest benefits of the technology.	New tellers	1982	Director: Tom Volotta Writer: Nicholas V. Iuppa Programmer: Bill Infield Disc manufacturers: Pioneer and Sony Production House: Bank of America, in-house facility
Customer Information Center	An interactive video information center designed to demonstrate bank services, qualify customers for them, overcome objections and close the sale. Includes highly visual sales sequences, printouts of recommendations and touch screen computation plans.	Bank of America customers for high ticket items	1983	Production team assigned after my departure.

2

Analysis and Instructional Design: A Quick Review

The first step in producing any type of interactive video program is to determine your objectives. This is a critical step since, no matter how well-produced the final program is, it may miss the mark if you haven't figured out what you are trying to accomplish.

Determining objectives is best done through several kinds of analyses which focus on performance problems and solutions. While these techniques are associated with training applications, rather than with selling or game playing, there are excellent parallels among all three. In all cases, a careful study of the desired behavior leads to a more successful program. This chapter offers a brief outline of the most important principles of instructional analysis.

For those of you who are well-acquainted with educational technology, please consider it a very brief refresher. For those just coming into contact with the science, this chapter represents the tip of the iceberg, a brief indication of how much more there is to learn.

KINDS OF ANALYSIS

Table 2.1 lists the three major kinds of analysis—front end analysis,[1] task analysis

---

[1] The term "front end analysis" was originally coined by Joe Harless in *An Ounce of Analysis (Is Worth a Pound of Objectives)* (Atlanta, GA: Guilde V Publications, 1972).

and learning problem analysis. It also shows the purpose, practical use and outcome of each. These different kinds of analyses will help you determine: 1. the cause of performance problems; 2. what must be done to correct them; and 3. the best technique to provide required training. Let's look at each area in more detail.

Table 2.1: Major Kinds of Analysis

Kind of Analysis	Purpose	Uses Related to Interactive	Product
Front end analysis (FEA)	To find the cause of a performance problem.	To determine if a proposed training program will be beneficial.	Report which should lead to an environmental, motivational or skill/ knowledge solution.
Task analysis	To define the skills to be taught in a training program.	To ensure that the training program includes everything necessary to perform the job.	Design document (instructional design).
Learning problem analysis	To identify learning problems of a specific nature.	To make sure that the proper teaching strategy is employed.	Worksheets, design document.

Front End Analysis

The front end analysis will determine whether the interactive program you plan to produce is the best solution to the problem at hand. For example, before you decide to produce an instructional video program to train people to do something better, it is wise to make sure that training will achieve the required result. There are, after all, several reasons why people are not doing what you want them to. Training assumes that people are not doing a job because they do not know how to.

There's an acid test for that theory. Go to several people who are doing the job incorrectly and "hold a gun," so to speak, to each person's head. Then insist that they perform the task. If they *can* perform correctly then you don't have a training problem. Perhaps the job isn't getting done because people don't want to do it; i.e., they aren't motivated. The solution to motivational problems is not more training: money and other incentives are more effective. But what if you give several people an ultimatum and they still cannot do the job properly? It still may not be a case of not knowing how; they may not have the right tools. The cause of this performance problem is environmental, not lack of know-how. Table 2.2 summarizes the three different causes of performance problems.

The reason for emphasizing the importance of front end analysis here is because you will be in deep trouble if your program doesn't work—and it won't work if you're trying to train people to do something they already know but don't want to do or

Table 2.2: Causes of Performance Problems

Kind of Performance Problem	Reason People Don't Do The Job	Solution
Motivational	Don't want to	Incentives
Environmental	Don't have the tools	Make a change in the environment
Skill/knowledge	Don't know how	Training

don't have the tools for. I suggest the first question to ask at the start of any big training project is "Has anyone looked into motivational or environmental factors relating to this problem?" If you alert your client to the possibility of other causes for the performance problem you will, at least, have some defense if your interactive video program doesn't deliver the kind of performance improvement your client expects.

Task Analysis

Once you have determined that training is necessary, you must pinpoint the skills to be taught. This is the purpose of task analysis. Task analysis is research, and unless you are in the most segregated of development groups you, as the instructional designer or writer, will have to do it. Certainly you will have to read all the background material you can find. You will also have to read procedures manuals and talk to all available experts. However, the strategic moment in task analysis comes when you sit down with the person who does the job just the way management wants it done, and you list every step that person follows, including tricks-of-the-trade. It's critical that nothing is left out and it's especially important that you identify the steps that are easily overlooked, but that make the difference between right and wrong. Those key steps are called *critical discriminations*.

For example, let's say I have designed a lesson on how to splice film. The steps in making a splice are: 1. place film in splicer; 2. cut off torn ends; 3. scrape off emulsion; 4. apply glue; 5. close splicer; 6. wait one minute. Which of those steps has a critical discrimination in it? Actually, it's step 4, applying glue. The other steps are fairly easy to teach. How much glue to put on is critical and rather hard to determine. You can only find out by spending a great deal of time talking to, even watching, someone who is an expert at film splicing.

Task analysis has a direct impact on the quality of the program. Before you conclude your analysis, ask yourself these questions:

- Have I really identified all the steps needed to do the job?

- Have I found all the secrets?

- Have I figured out where and how to explain them?

The point here is that if you don't do this job well the rest of the design team will be building a Cadillac without an engine. If you are going to work too hard on any aspect of the program, work too hard on this!

Behavioral Objectives

When you complete your task analysis you should list your goals as concrete *behavioral objectives*. Behavioral objectives are the goals of the program stated in terms of *action:* how people's behavior will change as a result of the program. Since behavioral changes are observable, they can be tested through pretest and post-test techniques. Testing is one of the greatest benefits of interactive video. There is a simple formula you can use to determine whether or not an objective is behavioral. Put it at the end of this phrase: "The student will be able to _____."
The phrase *will be able to* calls for a specific action verb or verbal phrase. Objectives like "learn math" don't work well. "The student will be able to *add*" is better. "The student will be able to *add all whole numbers in columns of 10 or less with accuracy of 90%*" is still better. It's very long, but very specific and therefore, easy to test.

It is very time-consuming to list all the things you want people to learn in terms that let you measure whether or not they have learned; however, it is necessary. Unfortunately, when everything must be figured out in advance, we lose those wonderfully spontaneous moments of instruction. Keep in mind, though, that for every wonderfully spontaneous instructor there were 500 instructors who rambled in the name of spontaneity. Happily, we lose them too.

The more specific you make each objective the better off you are. But what do you do with general objectives, such as "the student will understand math"? Most instructional designers feel these types of objectives are too general. I have always thought they were good to include in sales documents for nontechnical people. General objectives such as "understands," "appreciates," etc. are fine for sales purposes. But to create a working design for an interactive video program you will need pounds of detailed measurable objectives.

What about teaching skills such as getting along with people or managing them? The "how to" part is so flexible that the temptation is to set fuzzy, general objectives that lead nowhere. The real answer, however, is to define an approach that is practical and workable. This is what we did when preparing our "People Skills" program at Bank of America. We selected one clear-cut approach to dealing with customers and taught "how to do it" by demonstrating that approach. Through a series of pretests and post-tests, opinion questionnaires and instructor interviews, it became clear that the technique worked.

The Subject Matter Expert

If your program is about a simple subject, such as basic addition, you undoubtedly will be able to figure out all the objectives yourself. Otherwise, you will need the help of a subject matter expert (SME). The SME enters the interactive video process at this objective-setting stage and stays with you right through to completion of production. The SME is an important member of the team and one you should work closely with: he or she may be the one to tell you that the form in scene six is obsolete.

Learning Problem Analysis

Once you know what to teach, the next trick is to decide how to teach it. This is generally the area in which people forget to be creative, original or skillful. They rely on a few pat formulas and the whole program fails.

Proper instruction requires selection of the right teaching strategies. To do this you must analyze the material you will be teaching and identify places where general strategies or special strategies apply. Joe Harless, in his Performance Problem-Solving Workshop material, lists general teaching strategies and several special strategies. The general strategies are summed up in Table 2.3. These strategies should be used in the design of all instructional interactive video programs.

Harless' special strategies relate to techniques for teaching sequence problems, generalization problems, discrimination problems and psychomotor problems. These are unique learning problems which may or may not apply to your audience. Table 2.4

Table 2.3: General Teaching Strategies*

Strategy	Definition	Useful Hints
Step size	The amount of information to be presented before practice exercises begin.	Shouldn't be too much (two or three ideas).
Sequence	The order of presentation.	Not always chronological; affected by any special strategies used.
Simulation	How realistically the behavior to be learned will be presented to the student.	Start without distracting details. Add details.
Feedback	How the student will get information on responses made.	Should follow responses as closely as possible.
Exercise	How the behavior will be presented, prompted and tested.	Should simulate the skill to be learned.
Practice	Can be *isolated* (practice without distractions) or *integrated* (practices within real-world context).	Repeat exercises, adding more and more realistic distractions.
Prerequisites	Fundamental or underlying concepts, language, consequences, etc.	Start by explaining why; set context.

*Adapted from Joe Harless' Performance Problem-Solving Workshop material. Used with permission.

identifies four very important special strategies. (Further explanation of the kinds of learning problems and the solutions related to them is provided in Chapter 4.)

The basic steps making up the analysis portion of the interactive video process can be summarized as follows:

- Look at the performance problem and make sure that lack of skill is the cause, and training is the solution.

- Figure out what you want to teach and write your goals as concise *behavioral* objectives.

- Analyze the skill you want to teach and uncover all the secrets that really make a difference in doing it right.

- Identify all the special learning problems that might effect the order in which you teach things.

- Outline the lesson sequence and determine how much to present in each step.

- Determine how to let the learners *practice* each skill.

- Determine what critical information must be tested to *make sure* the student knows the information.

The next step is to use all of this information to prepare a design document.

Table 2.4: Special Teaching Strategies

Strategy	Definition	Useful Hints
Shaping	Presenting information in successive levels of difficulty.	For fine discriminations of complex behavior.
Grouping	Teaching similar things together to point out the difference.	Used when two or more items are so similar that distinguishing between them becomes difficult.
Mediation	A memory aid, rhyme, etc.	Used to strengthen relationships between elements when no natural connection exists.
Chaining	Presenting the last step first, second to last next, etc., but requiring their performance in order.	Used when there are many steps or when there is danger of the steps being performed out of sequence.

PREPARING A DESIGN DOCUMENT

A design document often has two parts: a client-oriented design summary followed by a detailed design outline. The outline is aimed at the developers and writers, but it is often presented to the client to illustrate the depth of the design. The design document is prepared after you have done a front end analysis and issued several status reports, so the client is already sold on the ideas presented.

Design Summary

The purpose of the client-oriented design summary is to make a case for the instructional strategies to be used and to confirm that progress is being made. This is often the place where program designers go into extended treatises on their chosen solution. As a general rule, avoid padding the design summary with lengthy descriptions of the benefits of interactive video or the advantages of one particular strategy over others. Keep your rationale objective. Don't leave anything out, but please avoid excesses.

If you work with an established design firm or any corporation's internal media department they will probably have a summary format or at least a good example of what has worked before. Follow it!

Design Outline

Preparing a design outline is rigorous work, but it's worth it. The design outline will serve as a roadmap for the program. At this point, you have completed discussions with your SME, dissected the content, prepared task lists, watched skilled practitioners doing the job and determined how management wants the job done. Now you have to organize the material into instructional sequences, and arrange the sequences into proper order by applying some of the techniques described earlier in this chapter. You also have to provide details for each of the sequences using your task lists, notes and observations as sources.

Next comes a detailed description of each item in the instructional sequence. If your company has worksheets for this process, by all means use them. At Bank of America, Pat Monteau of the Training Services unit created her own worksheet for the People Skills program. Figure 2.1 shows each major segment of the worksheet.

Item number: a breakdown of the instructional sequences—usually a number followed by a letter followed by a number, etc. The letter-number "combo" allows you to identify parts of the same item, such as different feedback sequences relating to the same exercise.

Kind of item: The major kinds of items include demonstration, exercise question, correct feedback, incorrect feedback, etc. These are explained in Chapter 4.

Figure 2.1: Major Segments of the Design Worksheet

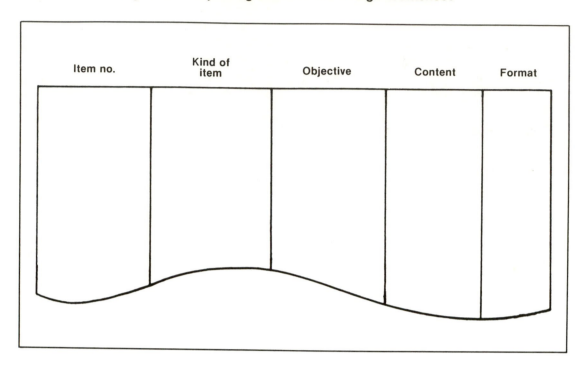

Objective: Tells the purpose of each item. Usually the objective only applies to each full item, e.g., Item 1, rather than the parts of the item; 1.A., 1.B., etc.

Content: A complete statement of the content presented in each part of an item. It can be an outline of the kind of activity taking place, or suggested words for the item itself. For example, ''Customer: 'These bills are kind of wrinkled.' Teller: 'You mean you want all new money.'''

Format: The way the content is presented, for example, still frame, motion sequence, motion followed by still frame, etc.

Figure 2.2 shows an example of the worksheet page with appropriate design information inserted under the proper headings. (The part of the design outline used in this example came well into the lesson and had to do with techniques tellers could use to clarify the customers' comments.) Remember the purpose of the design outline is to communicate the organization and instructional strategies you've come up with. If you don't do it accurately or completely, the writer and director will have little or no instructional plan or, worse, they will make up their own.

The next step is to put all this information into a form that program designers, computer programmers, etc., can follow. Therefore, Chapter 3 will examine flow-charting, an essential component in the interactive video design process.

Figure 2.2: Completed Design Worksheet Page

PROGRAM: _____ SEQUENCE __5__ PAGE __2__ of __17__

Item No.	Kind of Item	Objective	Content	Format
5.D	Question 1	The student will be able to recognize the act of para-phrasing a customer statement	Customer: (says something) Teller: (paraphrases customer's words)	Motion sequence followed by freeze frame with choices: 1. Paraphrasing 2. Amplifying 3. Non-clarifying
5.D.1	Repeat question	Same as 5.D	If you would like to hear the question again press _____.	Menu
5.D.2	Feedback: incorrect	Same as 5.D	You didn't get it right.	Feedback frame
5.D.3	Feedback: correct	Same as 5.D	Correct. The teller paraphrased what the customer said.	Freeze frame, high-light
5.D.4	Menu: incorrect	Same as 5.D	If you would like to: -repeat the question, press _____. -obtain more instruc-tion, press _____. -go to the next ques-tion, press _____.	Menu
5.E	Question 2	Student will be able to recognize dis-agreement as non-clarifying.	Customer: (says something) Teller: (disagrees)	Motion sequence followed by freeze frame with choices: 1. Paraphrasing 2. Amplifying 3. Non-clarifying

3

Flowcharting and Program Design

FLOWCHART NOTATION

Flowcharting is an integral part of the interactive design process. It is done to communicate sequence, decision points, branching and the flow of information in an interactive video program. In traditional linear video programming there is no need to document the various paths a viewer can take through the program, since there is only one path. In interactive video, flowcharts not only show all possible paths a viewer may take, but also indicate the various activities that will take place: video motion sequences, subroutines, tests, etc.

Each element of an interactive video program is represented by a particular symbol. Figure 3.1 shows the major interactive video flowcharting symbols and offers a brief description of each.

LEVELS OF FLOWCHARTS

Flowcharts are used for a number of different purposes. They are used to communicate general program flow to clients and writers, to indicate the organization of all parts of the program to production staff, and to indicate the mechanics of the program to computer programmers. Since the needs of one group are quite different from those of another, there are three levels of flowcharts, each designed for a specific purpose. Table 3.1 identifies the three levels and their appropriate uses.

Figure 3.1: Interactive Flowchart Symbols

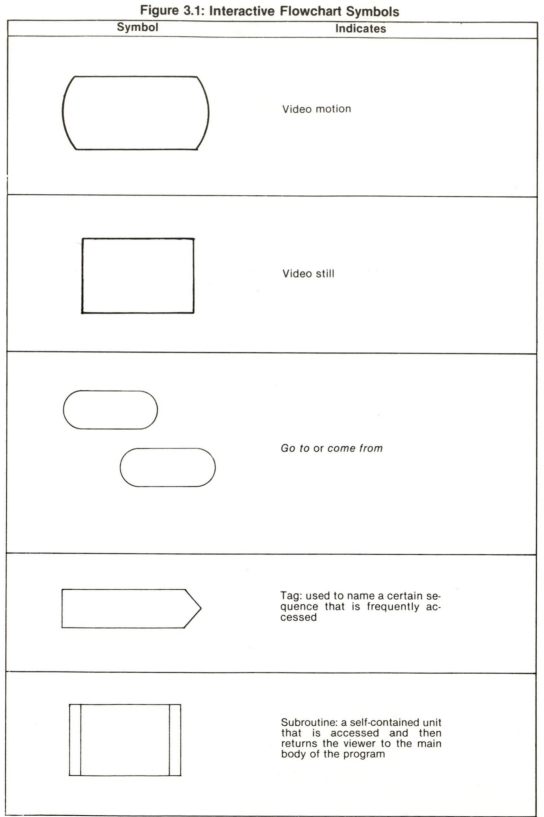

Symbol	Indicates
	Video motion
	Video still
	Go to or *come from*
	Tag: used to name a certain sequence that is frequently accessed
	Subroutine: a self-contained unit that is accessed and then returns the viewer to the main body of the program

Figure 3.1: Interactive Flowchart Symbols (cont.)

Symbol	Indicates
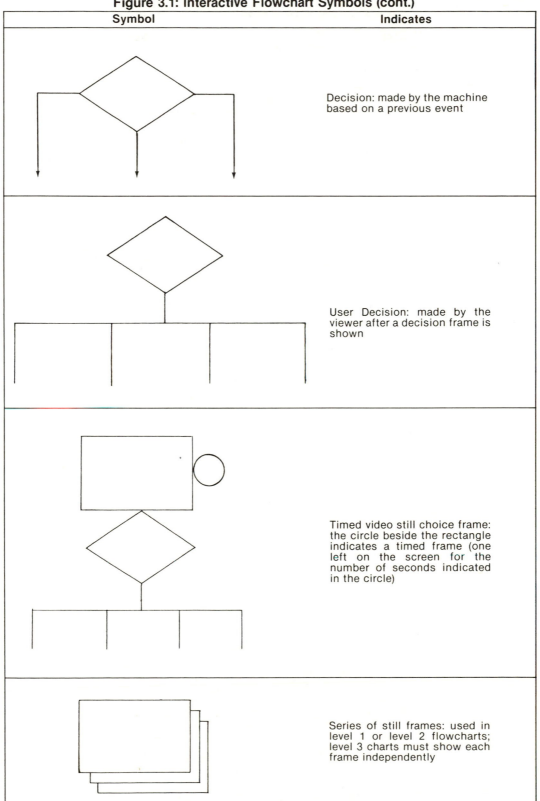	Decision: made by the machine based on a previous event
	User Decision: made by the viewer after a decision frame is shown
	Timed video still choice frame: the circle beside the rectangle indicates a timed frame (one left on the screen for the number of seconds indicated in the circle)
	Series of still frames: used in level 1 or level 2 flowcharts; level 3 charts must show each frame independently

Figure 3.1: Interactive Flowchart Symbols (cont.)

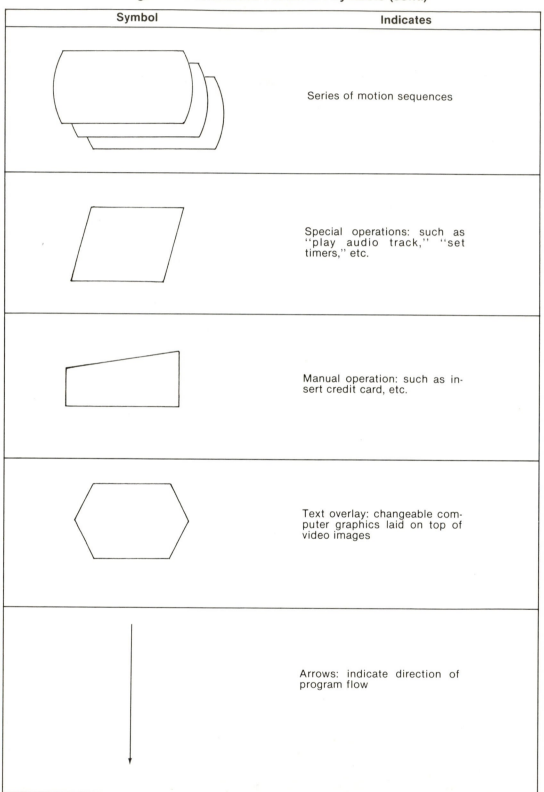

Symbol	Indicates
	Series of motion sequences
	Special operations: such as "play audio track," "set timers," etc.
	Manual operation: such as insert credit card, etc.
	Text overlay: changeable computer graphics laid on top of video images
	Arrows: indicate direction of program flow

Figure 3.1: Interactive Flowchart Symbols (cont.)

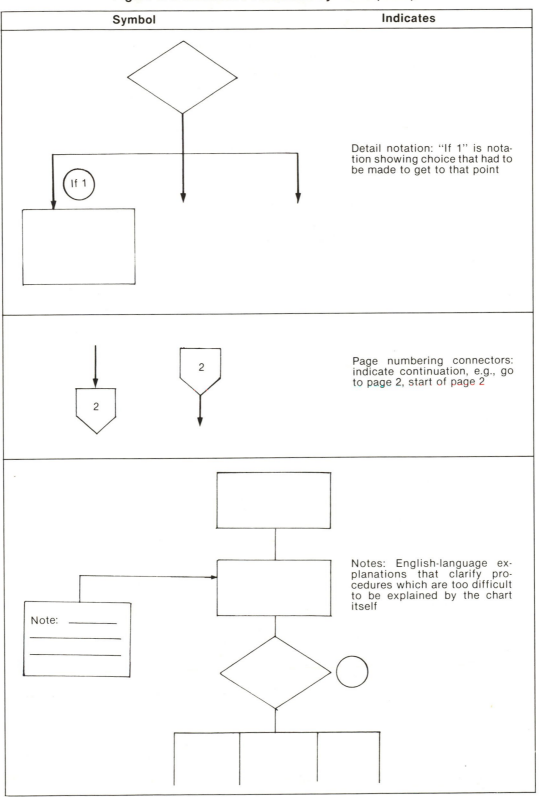

Symbol	Indicates
	Detail notation: "If 1" is notation showing choice that had to be made to get to that point
	Page numbering connectors: indicate continuation, e.g., go to page 2, start of page 2
	Notes: English-language explanations that clarify procedures which are too difficult to be explained by the chart itself

Table 3.1: Levels of Flowcharts

Flowchart Levels	Purpose
1	To communicate general program flow to clients and writers prior to scripting stage.
2	To indicate complete organization of all parts of the program to production staff and decision makers.
3	To indicate the mechanics of the program to computer programmers.

Level 1 flowcharts document the paths available to the user or viewer and are based on the information developed in the design document. Level 1 flowcharts are frequently used to ''sell'' the interactive program to clients. For this reason, they are often illustrated and mounted on large presentation boards. A good set of level 1 flowcharts can accompany the design documents.

Level 2 flowcharts add in all the details of the program, but leave out the mechancal instructions needed by computer programmers. Level 3 flowcharts contain all the program details, mechanical instructions, counter instructions, frame numbers, etc., that will be programmed into the computer. It takes intense concentration to develop level 3 charts. However, the effort is necessary because the program won't work if level 3 charts are incorrect. When you are working with the on-board microprocessor of an industrial disc player (where the computer program is encoded onto the video disc), errors in the flowchart can lead to reprogramming, remastering and, consequently, thousands of dollars of added expense.

Designing flowcharts requires three basic characteristics: a logical mind, tremendous patience and strict attention to detail. The general process is to identify each content element, label each element and connect the elements into a logical flow. Then, you should check and recheck to see that all necessary details, including loops, computer instructions, etc., are present. Level 1 and level 2 flowcharts are often done before scripting begins. Level 3 charts are usually completed during or *after* shooting so that refinements in the script can be incorporated into the program.

Figures 3.2, 3.3 and 3.4 show level 1, 2 and 3 flowcharts, using the flowchart notation described earlier. Each chart combines the symbols into a unified sequence. All three charts present the same sequence with increasing levels of detail.

Figure 3.2 presents a level 1 flowchart of a ''point-of-sale'' disc, featuring the menu which will lead to one of several product stills. The still, in turn, leads the viewer to a choice of activities relating to the product.

In the level 2 chart shown in Figure 3.3, terms are used to identify key blocks of information. Words that are repeated are abbreviated: for example, the phrase ''Product Still'' is abbreviated PS. The flowchart makes all parts of the program clear and

Figure 3.2: Level 1 Flowchart for Point-of-Sale Disc

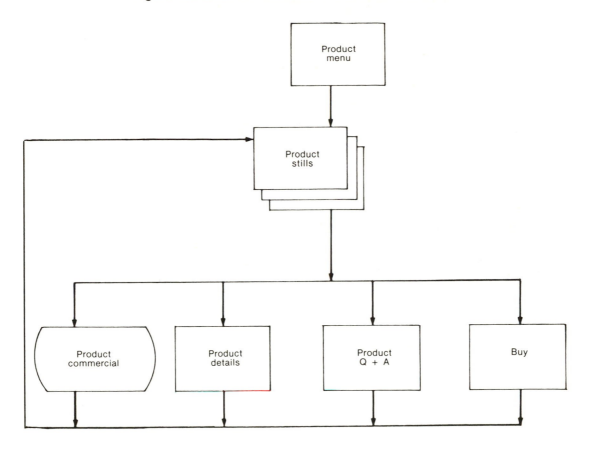

can be used by production personnel to begin work on the video and other sequences. However, level 2 charts cannot be used by computer programmers.

The level 3 flowchart shown in Figure 3.4 is so detailed that it must be done in code to get all the information on the page. The amount of detail makes any superficial assessment of the chart's accuracy impossible—it must be worked through step by step. Using this chart, the programmer can put an acceptable computer program together. Despite the amount of detail present this flowchart is still incomplete. Frame numbers must be added if the disc is to be programmed accurately, but this cannot be done until editing—and sometimes disc replication—is complete.

The Gantt Code

The level 3 flowchart shown in Figure 3.4 uses a variation of the code invented by Rodger Gantt, a Bank of America instructional designer. The Gantt Code, as we call it, uses activity names or code numbers to represent all elements of an interactive video program. While the Gantt Code was constructed for instructional material, the principle could apply to other types of interactive video programs as well.

Figure 3.3: Level 2 Flowchart for Point-of-Sale Disc

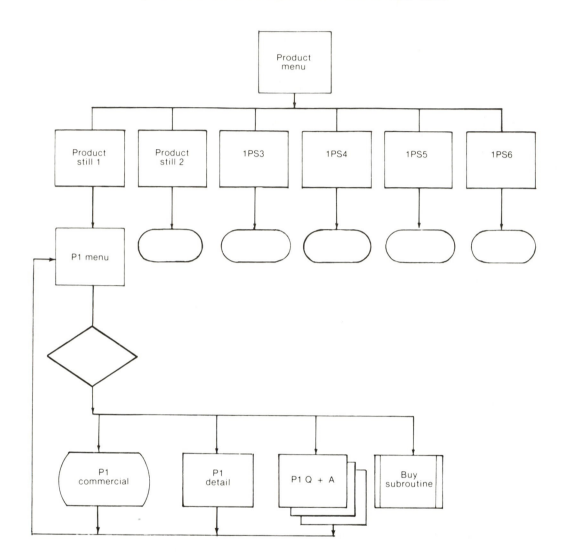

In Gantt's system, the activities are named, abbreviated and numbered. The main purpose of the code is to create a clearcut shorthand that designers and programmers can use to identify activities. It saves writing time and provides for greater consistency when program elements are being discussed. Each element of the code is explained below.

The major building blocks of an instructional interactive video program are demonstrations, exercise questions and test questions. These are represented in the Gantt Code as follows:

Demonstration	=	DEMO
Exercise question	=	XQ
Test question	=	TQ

Figure 3.4: Level 3 Flowchart for Point-of-Sale Disc

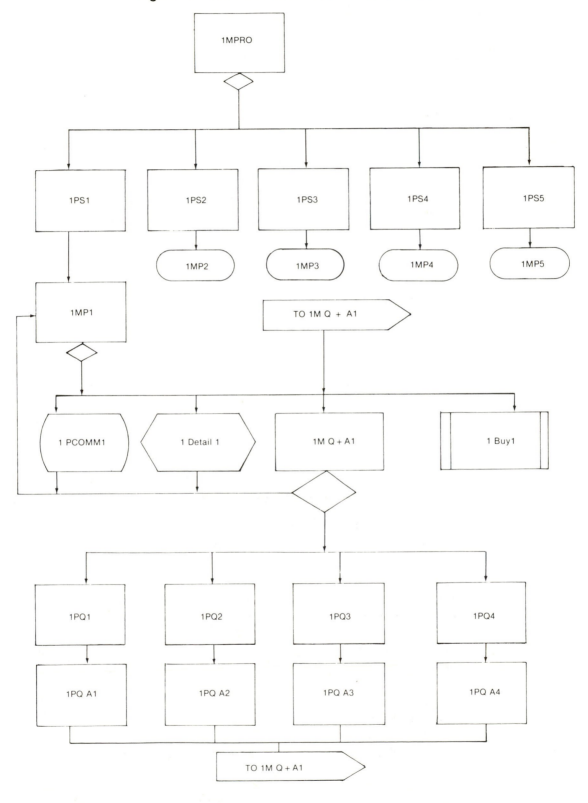

Other blocks include introductions, reviews, previews and various types of menus:

Introduction	=	INTRO
Review	=	REV
Preview	=	PRE
Test pass menu	=	MTP
Test fail menu	=	MTF
Review menu	=	MREV
Preview menu	=	MPRE

In addition to exercise questions (XQ), there are three different kinds of exercise feedback: correct, incorrect and common feedback (a single response that works for all answers, right or wrong). There are also drills with questions and feedback:

Drill questions	=	DQ
Drill feedback	=	DF
Exercise questions	=	XQ
Correct feedback	=	XCF
Incorrect (false) feedback	=	XFF
Common feedback	=	XCOM

In programs with more than one lesson, a number usually precedes the item, to indicate the lesson:

Lesson 1, demonstration	=	1DEMO
Lesson 1, review	=	1REV
Lesson 1, exercise question	=	1XQ

Since there can be several activities of any kind in any lesson, the activities are given numbers which are placed *after* the lesson number and item code:

Lesson 1, demonstration 1	=	1DEMO1
Lesson 1, demonstration 2	=	1DEMO2
Lesson 1, demonstration 3	=	1DEMO3
Lesson 2, exercise question 4	=	2XQ4
Lesson 2, correct feedback for exercise question 5	=	2XCF5

Similarly, since there can be more than one incorrect (false) feedback to each question, the number of the feedback is indicated by a point after the activity number:

Lesson 2, exercise question 5, incorrect feedback 1	=	2XFF5.1
Lesson 2, exercise question 5, incorrect feedback 2	=	2XFF5.2

Figure 3.5: Feedback Segment of Flowchart in Code

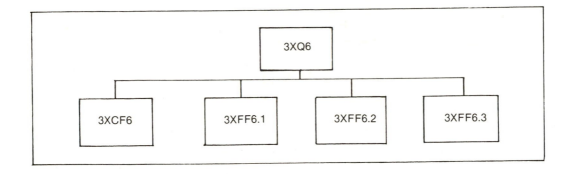

In flowcharts, those numbers would also be used to indicate different feedback corresponding to choices in a multiple-choice question. For example, Figure 3.5 shows the sixth exercise question in lesson 3, where there are three incorrect possibilities and one correct choice.

Creating Your Own Code

A final note on flowchart coding. You can create your own identifiers for interactive game programs and point-of-sale as we did in Figures 3.2, 3.3 and 3.4., as well as training. Table 3.2 gives a few suggestions.

Table 3.2: Point-of-Sale and Game Program Codes

Attract Mode	=	AM
Welcome Sequence	=	WEL
Learn Menu	=	ML
Learn Demo 1	=	LD1
Learn Demo 2	=	LD2
Product Showcase Menu	=	MPS
Product Showcase 1	=	PS1
Product Still 1	=	PStill1
Product Detail 1	=	PD1

Thus far, we have discussed the three levels of flowcharts used to communicate information about the organization of the interactive video program to various clients, scriptwriters, production personnel and programmers. In the remainder of this chapter, we will examine the basic design principles used to create and organize the major building blocks of instructional interactive video program lessons and tests.

BASIC LESSON STRUCTURE

The fundamental elements of instructional design are lessons and tests. In interactive video, a lesson is an entire unit of information on a single subject or on very few related subjects. For example, in banking the subject of "debits and credits" is taught as a single lesson, since the two subjects are so closely related. In auto mechanics, "operations of the combustion chamber" would be a single lesson. However, "the complete operation of the internal combustion engine" would obviously be too complex to present in one lesson.

Lessons include demonstrations and exercises. As mentioned earlier, a demonstration explains what something is or how to do it; an exercise lets you practice doing it. Tests are given to find out if you have learned from the lesson. Technically, they are not part of the lesson.

Figures 3.6 and 3.7 present two basic lesson design formulas. Keep in mind that they show the flow of information in a *lesson;* they are not meant to be universal formulas for *exercises.* The design of exercises is governed by content and objectives, as discussed in Chapter 4.

The basic lesson design flowchart in Figure 3.6 indicates that you demonstrate something and then ask one or more questions about it in a test. If viewers answer the questions correctly, they go on to the next lesson. If viewers fail the test, the lesson and test are repeated.

Figure 3.6: Basic Lesson Design Flowchart

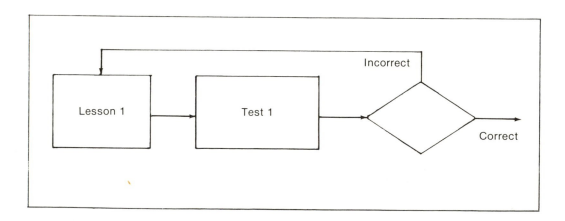

Figure 3.7: Expanded Lesson Design Flowchart

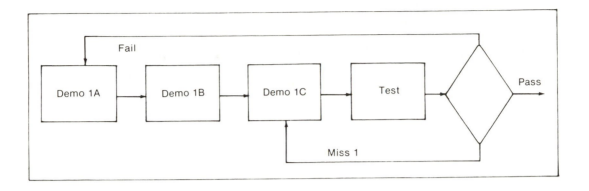

Figure 3.7 represents a modest expansion of the design shown in Figure 3.6. In Figure 3.7 there are several demonstrations and a multifaceted test. If viewers pass, they go on to the next lesson. If they miss several questions, they start over at the beginning of the first demonstration. If they miss one question, they go back to the part of the lesson that deals with whatever misconception led them to their error (in Figure 3.7, this would be the material presented in Demo 1C). Important details to note are that lessons include both demonstrations and exercises; tests have many questions.

All major kinds of lessons, then, can be summarized with one flowchart, shown in Figure 3.8. However, an interactive training program usually consists of more than one lesson. Plus, there are other activities that must be carried out as part of a total training program. Let's move beyond lessons and tests to see how lessons fit together into whole training programs. We will also consider elements that can be added to the overall program design to make it more effective.

Figure 3.8: Universal Lesson Design Flowchart

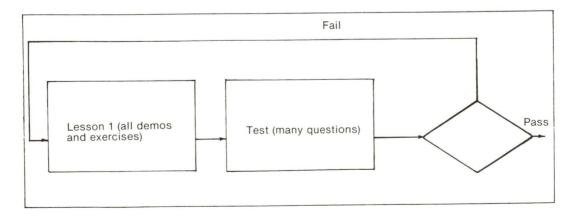

Figure 3.9: Pre-Determined Lesson Order

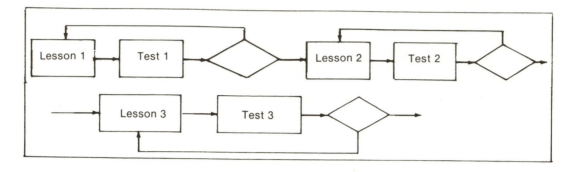

Lesson Order

As we know, interactive video programs are usually made up of several lessons. In the following discussion, we will present examples of programs that consist of three lessons each. Lessons can be set up in unchangeable order (see Figure 3.9). In that case, the viewer must proceed from one lesson to the next. However, lessons can also be designed so that the viewer can pick the lesson order using a menu (see Figure 3.10). Finally, lessons can be arranged so that the viewer must complete a prerequisite lesson before choosing between other lessons (see Figure 3.11).

Introductions and Exams

In addition to lessons and tests, most instructional designs call for introductions, which include objectives, explanations and overviews. There are also final exams which integrate information from *all* lessons.

Figure 3.10: Viewer-Determined Lesson Order

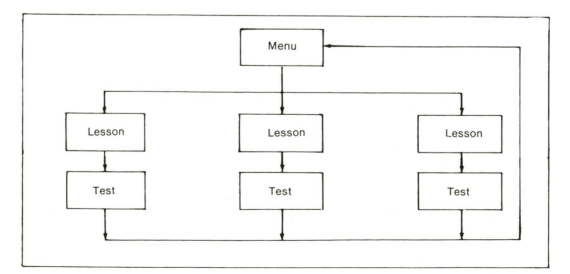

Figure 3.11: Viewer-Determined Order with Prerequisite Lesson

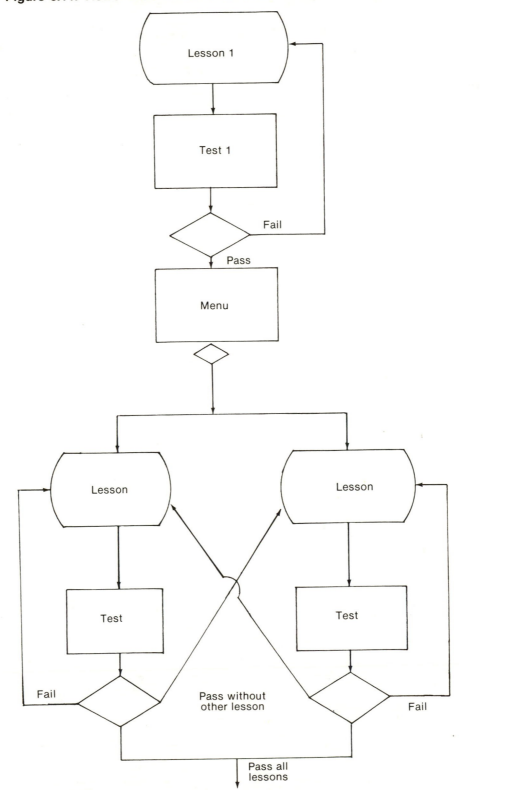

Figure 3.12: Master Flowchart for a Simple Instructional Program

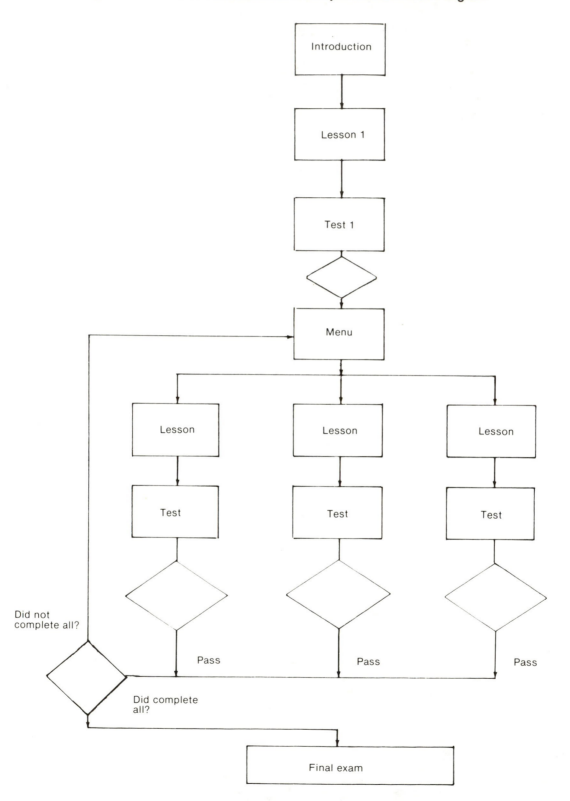

Figure 3.13: Lesson Review Technique

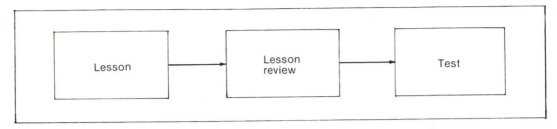

A master flowchart for a fairly simple instructional program, integrating all of these elements, would look like the one shown in Figure 3.12. Note that the test fail path is not shown. This omission is acceptable in a level 1 flowchart because it is obvious, however level 2 and level 3 flowcharts would require it.

Enhancing Basic Designs

There are various techniques that can be used to add some sophistication to lesson designs, and make them more effective. These include lesson and program reviews, pretests, and even prologues and epilogues to help with storytelling.

The flowchart in Figure 3.13 shows how you can insert a review in each lesson before the test segment. The design in Figure 3.14 lets the viewer review the whole pro-

Figure 3.14: Program Review Technique

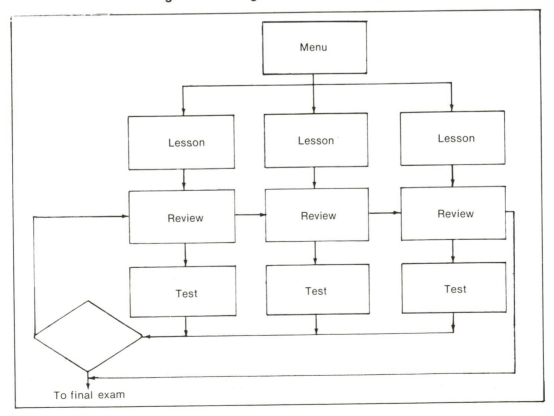

gram by skipping from one review segment to another. After completing the test, the user jumps back to a review and then immediately branches to the other reviews before taking the final exam.

Figure 3.15 shows how the instructional design can allow for a pretest that permits the learner to skip certain lessons. Here, once the viewer passes the pretest, he or she skips lessons 1, 2 and 3, and proceeds directly to 4.

A fully developed program design flowchart incorporating all of these techniques is shown in Figure 3.16. What should be clear is that, depending on the design tech-

Figure 3.15: Pretest Technique

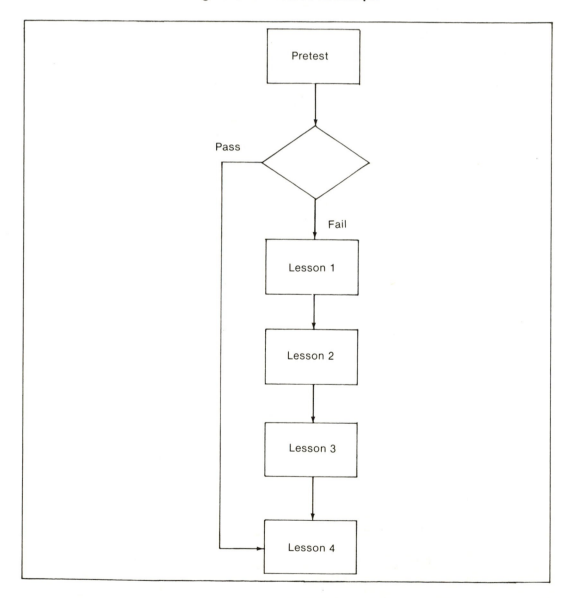

Figure 3.16: Fully Developed Master Flowchart

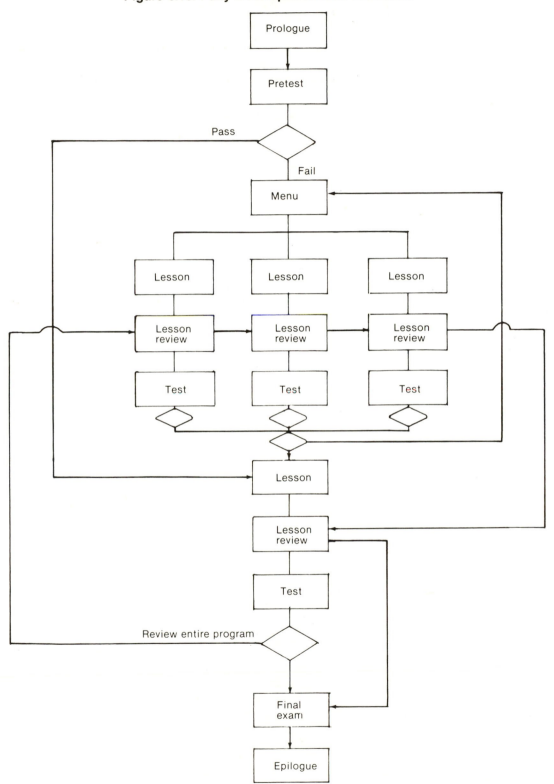

niques you use, you can create one instructional video program that will serve a variety of trainees who are at different entry levels.

The flowchart and instructional design techniques presented in this chapter will help you build the framework for your own interactive video programs. In Chapters 4 and 5, we will examine the design strategies that influence how effectively your program will meet its objectives. These strategies pertain to the design and construction of exercise and test questions—the heart of an interactive program.

4

Exercise Design and Evaluation

Interactive video exercises are most effective if they are simulations of the behavior to be learned. Since simulation techniques change depending on the kind of learning problem, there can be no universal exercise formula. In this chapter, we will look at several design formulas, based on different kinds of learning problems.

BASIC DESIGNS

Discrimination Learning Problems

Discrimination is the ability to differentiate between several items. For example, at Bank of America our "Debits and Credits" video program taught student tellers how to tell the difference between three kinds of stamps. The tellers had to be able to recognize when to stamp a customer's check or deposit slip with a "batch" stamp, an "interbranch" stamp or a "cash-paid" stamp. (Cash-paid is used only when cashing a check or paying on a withdrawal.) The logical exercise format to choose in this case was simple three-way multiple choice. Figure 4.1 diagrams the part of the lesson dealing with the cash-paid stamp.

As Figure 4.1 shows, upon picking one of three choices, students were given feedback tailored to the particular choice. If students selected the correct answer (choice 3), they got positive reinforcement—a pat on the back. There was even some elaboration on why the answer was right before the lesson continued. If students selected choices 1 or 2 (both incorrect), they got feedback explaining why the particular answer was wrong. The program then looped back, not to the start of the lesson, but to the *question.*

Figure 4.1: Simple Multiple-Choice

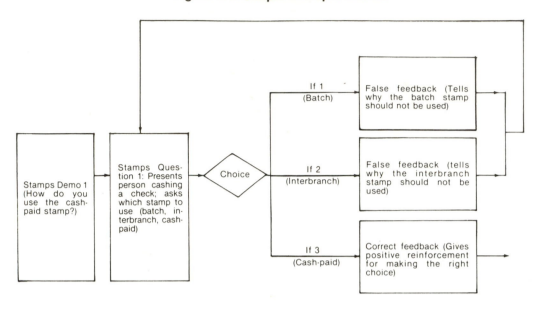

By using branching, you can design feedback tailored to the learners' responses. Using specially constructed feedback is far better than just repeating a section of the previous demonstration to provide feedback. Tailored feedback can go into detail about why the answer was wrong and can add new insight into the learning experience. Figure 4.2 shows how you can use branching to expand on the simple multiple-choice formula shown in Figure 4.1.

The branch occurs after the first question, where two diverse paths begin. If the student is right, he or she gets correct feedback and goes on. If the wrong answer is given, the student is asked a completely new question (XQ2). A correct answer to that question steers the student back into the mainstream. A wrong answer receives incorrect or false feedback (XFF2), etc.

Figure 4.2: Complex Multiple-Choice

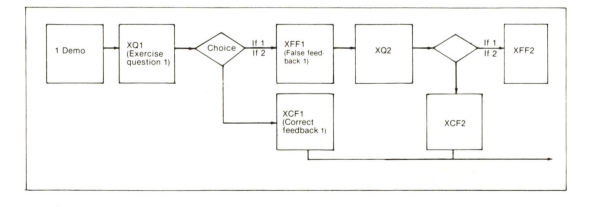

Figure 4.3: Common Feedback

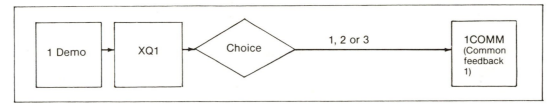

An alternative to the branching just described is common feedback. Common feedback is written in such a way that it explains both correct and incorrect answers. There is no branching or looping back following common feedback—students proceed to the next demonstration or lesson (see Figure 4.3). Common feedback is a technique that should be reserved for questions that are part of a series, such as minor variations on the same idea. Common feedback is less specific and therefore less effective than branched multiple-choice, but in many situations it is quite adequate.

Generalization Learning Problems

Generalization is the flip side of discrimination. Instead of learning to tell things apart, you learn to put them into groups. The most common type of exercise used to teach generalizations is an identification drill, in which a variety of items are presented and must be grouped into several categories.

In the "People Skills" program, students were presented with a series of statements and were asked to identify them either as statements that clarified the conversation or got off the point. In general, students first worked through exercises that helped establish the basic rules. Then, a number of drill questions (DQ) were presented, followed by common feedback for either answer as shown in Figure 4.4.

Figure 4.4: Identification Drill

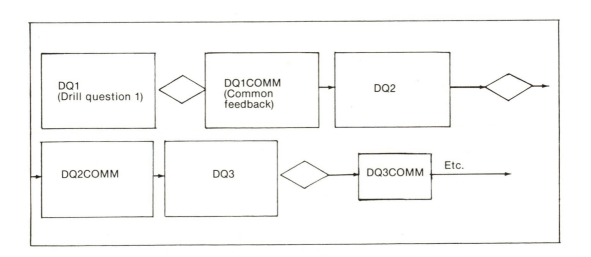

The exercises that teach the basic generalization concepts are as important as the drill. Again, a multiple-choice design works especially well, particularly if you use an inductive multiple-choice, in which the student draws upon previous knowledge to find a new concept. When constructing an inductive multiple-choice exercise, you should not present the exact information needed to answer a specific question; instead you allow the student to make an educated guess. Most students find this technique quite enjoyable. The following is an inductive question on which stamp to use, from the "Debits and Credits" program:

There are three teller stamps: the batch stamp used in many transactions, the cash paid stamp used only in cash-paid transactions, and the interbranch stamp used only in interbranch credits.

Which stamp do you use on credits accepted for *deposit* at *your* branch?

Even if you are unfamiliar with banking, your sense of logic may lead you to pick the batch stamp. This type of inductive exercise helps to clarify a principle and also helps students remember what they have learned. I highly recommend inductive exercises, especially when you are dealing with sophisticated audiences.

Sequence Learning Problems

Sequencing lessons teach how to get things in the right order. Many mechanical skills have a sequencing component. For example, changing a spark plug involves several steps that must be performed in the right order, or the job won't be done correctly, and the car won't work. The types of exercises used in sequencing lessons might include ordering exercises, in which the student rearranges tasks into the correct order. There is also a variation on fill-in-the-blanks, in which the viewer lists tasks in the right order.

In "People Skills," we had to teach the three steps in greeting a customer: recognize, identify and stroke. We presented students with this incorrect order of steps — 1. stroke, 2. recognize, 3. identify—and accepted as the correct answer their pressing numbers 2-3-1 in succession. The diagram in Figure 4.5 shows a simple way to construct this sequencing exercise.

Psychomotor Learning Problems

Using interactive video to teach pyschomotor skills has led to the most experimental application of the technology. Until now, very few people used interactive video to teach psychomotor skills because the lessons require the simulation of physical activities.

Perhaps one of the best examples of real psychomotor lessons can be seen on the "First National KIDISC," where hand-eye coordination is strengthened by having the

Figure 4.5: Sequencing Exercise

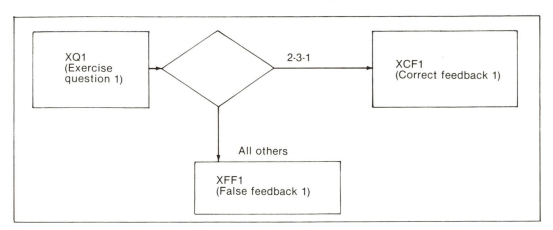

child press a button as a series of frames flash by. The sequence is shown in Figure 4.6. The frames move from a highlighted ring on the outside of a bull's-eye, closer and closer to a highlighted strike at the bull's-eye itself. If the child pushes the button too soon, he or she gets to see just how far off the selection was. A bull's-eye hit shows up with a gaudy congratulatory frame. The same principle can surely be applied directly to the popular outer space adventure/target games. However, a joystick or push-button controller may be needed.

Figure 4.6: Psychomotor Exercise

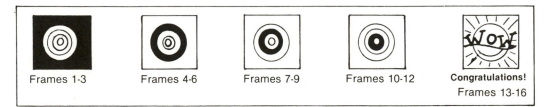

Interactive psychomotor exercises will work for driver training, pilot training, health education, pedestrian safety, map skills training, etc. They will probably also work well for high-simulation video games. Among the more notable examples of these applications are the Massachusetts Institute of Technology's Movie-Maps Project directed by Nicholas Negroponte, and the American Heart Association's CPR Training Program pioneered by David Hon.

All of the basic exercise construction techniques that I have described are summarized in Table 4.1. Next, let us examine more sophisticated variations of these exercises.

INTERMEDIATE DESIGNS

There are several kinds of interactive exercises that are variations or expansions on the basic formulas just described. Let's review each of them.

Table 4.1: Basic Exercise Techniques

To Teach	Use
Discrimination skills	Simple multiple-choice Complex multiple-choice Common feedback
Generalization skills	Simple multiple-choice Drills Inductive multiple-choice
Sequencing skills	Ordering exercises
Psychomotor skills	Simulation games Target games

Matching Exercises

To teach generalization skills, three to six items are presented which must then be matched to three or four other items or categories. The items matched can be individuals, groups or species to be identified by the appropriate name or category. For example, dates can be matched to events, kinds of sores or wounds can be matched to diseases, cars or trucks can be matched to model names. Matching is a realistic and valuable simulation technique.

Internal Matching and Rating

This technique can be used if you have an on-board microprocessor or external computer. The computer or microprocessor compares several ratings made by the student to determine the student's preference or attitude. For example, in "People Skills," we showed a customer/teller argument. Then we asked students to rate the behavior of the teller and the customer to determine which of the two they favored. The computer compared the two scores and branched to individual paths through the program based on the student's bias as expressed in their preference.

Rating/matching exercises are excellent techniques for segregating your audience by the preconceptions they bring into the program. These preconceptions can often block out information if they are not taken into consideration when teaching. Rating/matching exercises are also good discussion starters for interactive video programs that are used in classrooms or with groups. There is usually no right or wrong answer.

Visual Discriminations

This is a variation on a multiple-choice exercise used to teach discriminations. Essentially, several different objects are presented. The viewer is asked to identify certain parts of each object, or, to differentiate between the objects. These exercises allow you to take advantage of the excellent representations made possible by the visual realism of the disc.

For example, if you want to teach people the difference between similar-looking items or parts of the same item, present the items or parts side by side and ask the viewer to make a choice. An example from banking is the debit/credit discrimination. A student could be asked which of a dozen different rectangular bank forms viewed on the screen is a debit and which is a credit.

As mentioned in Chapter 1, motion should not be overlooked. While an interactive video disc probably can't teach you how to throw a forward pass or play the piano, it can demonstrate the correct procedures and teach people to recognize the right way from the wrong way.

Consequence Remediation

This may be one of the most effective types of interactive multiple-choice exercises. Instead of branching from the question to a teacher who tells you why something is right or wrong, or looping back to see a repeat of the lecture segment, consequence remediation shows you the *results* of your choice. It gets you to do things right by showing the negative or positive consequences of your actions. The following is an example of consequence remediation from "People Skills." The sequence is also diagrammed in Figure 4.7.

Customer: What do you mean you have to place a hold? Listen, I've been a customer of this bank for eight years. I've never bounced *one* check and if you don't approve this thing right now I'm withdrawing every cent.

Teller: (This response is choice number 2 of three choices) Don't talk to me that way, sir. If you really had been a customer for eight years you'd know that holds are a standard part of bank operations and we have to follow procedures. Period!

Customer: (Consequence remediation to choice number 2) Follow procedures. Well then fine, start following the procedures to close my accounts, period!

At this point the narrator can come back and offer a commentary on the transaction.

Narrator: You really blew it. By challenging the customer's authority, you made him feel as though he had to act his toughest. He may still back down though. Remember, we're trying to clarify the customer's feelings. Try to resolve the situation.

Figure 4.7: Consequence Remediation

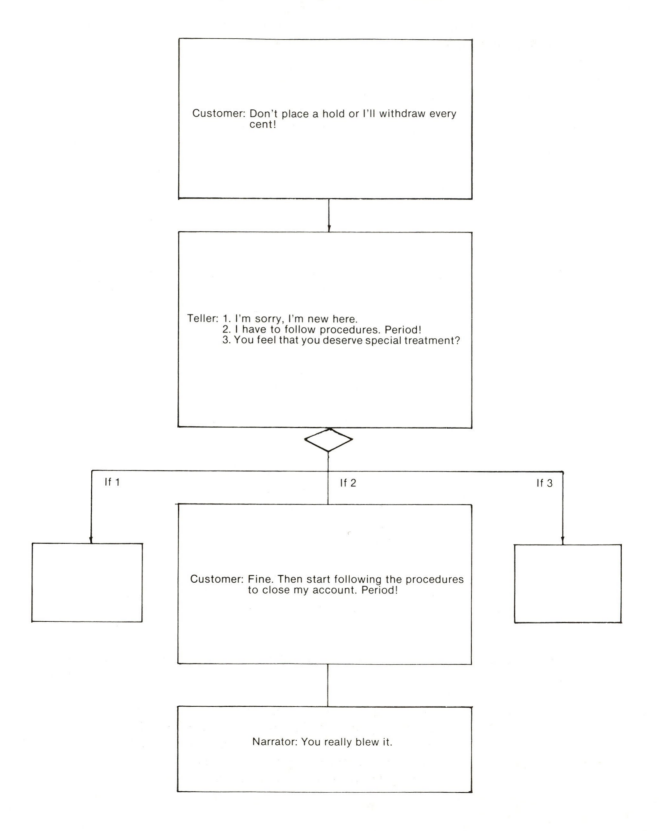

Consequence remediation is an extremely powerful tool. It is especially powerful when used in an interactive video program, since the medium can depict the consequence so realistically. It is an outstanding technique for teaching personal interactions or any procedure where improper choices lead to very pronounced consequences. Consequence remediation is also the underlying principle in many of the advanced exercise designs, which we will discuss in the following section.

ADVANCED EXERCISE DESIGNS

In this section we will consider more advanced types of interaction. As has probably become clear, the more advanced the exercise, usually the more branching is employed.

Spiderwebs

Branching generally leads people to think of the learning tree, in which one decision leads to two more choices, each of which leads to two more choices, etc. The problem with trees is that decision making, or human discussion, seldom works that way. Actually, there is a great deal of redundancy in decision making. In arguments or even friendly discussions, people keep returning to the same points. So, in the end, people are limited to a few logical conclusions.

Interactive video producers want to limit possible options, and rightly so. No one is really going to produce a tree that goes into infinity—it's impractical and impossible. More realistic and far less expensive is an exercise called the *spiderweb*. It operates on the same principle as the learning tree but differs in that certain responses are repeated, as they would be in real life.

In "People Skills" we wanted to show a loan rejection interview. We knew that the loan officers would keep coming back to the same points over and over again. Figure 4.8 presents a graph and summary of the beginning of the rejection interview.

You can see that despite the fact that there is a wide variety of choices, several responses keep reappearing—just as, I feel, they would in real life. The loan officer (Joe) has figured out what to say to the customer in advance. That's only logical, as is the idea that he would respond with the same or similar words in situations in which he was confronted with extreme anger.

Spiderwebs are an excellent technique for simulating a discussion. The viewer takes control of the responses of one party in the discussion and selects the best response to statements by the other person. There are several options for each response, and each response in turn generates a new statement. Because many choices lead to the same response, the exercise does not go on forever, but instead, eventually resolves itself in one or two logical outcomes.

A spiderweb takes three to 15 minutes for a viewer to work through, but it also takes a skilled instructional designer and a talented writer to make it work. If you are

Figure 4.8: Loan Interview Illustrating Spiderweb Design

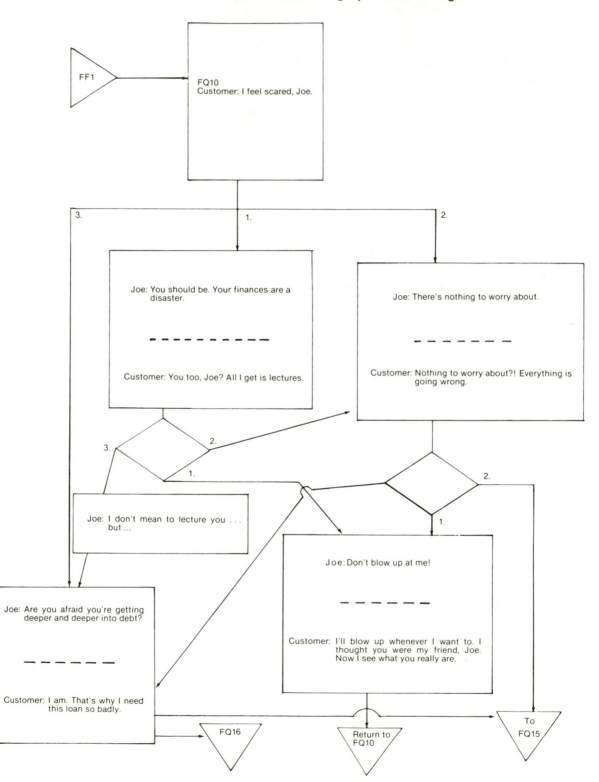

going to try spinning spiderwebs, make sure that you check to see what happens when you follow every path. Do they all connect? Do the right answers sound reasonable?

Invisible Spiderwebs

Unfortunately, in real life, conversations don't come to a screeching halt when you are faced with three or four well-defined choices. In real life, if you don't make a decision by a certain point, a conversation or situation still continues along the same path.

At present, spiderwebs are structured the way they are, to some degree, because of the characteristics of video disc players. Current disc set-ups usually require the viewer to indicate his or her choice during a freeze frame. There is no way to make decisions while the program plays on. To overcome this you can use an outboard computer, which can be programmed to open up a decision-making window. The viewer is informed at the outset that if he or she indicates a choice anytime during a window, the program will branch.

What kind of choices would you be making in one of these invisible spiderwebs? In an interpersonal relations course, you may just indicate positive, negative or neutral feelings. Referring to the previous example, let's say you are in a discussion with a loan officer and are asked to push certain buttons whenever you feel a certain way. If you do nothing the program proceeds in a neutral middle-of-the-road course. But if you indicate positive or negative feelings, the program branches, and the other person responds accordingly.

What I have just described is a program that does not come to a halt and wait for the viewer to make decisions; it responds to opinions that are indicated while it is going on. The flowchart for that program is a spiderweb, but since the questions (in this case the decision-making windows) aren't indicated to the viewer, it is indeed an *invisible* spiderweb.

HOW TO EVALUATE EXERCISES

In making decisions about the kind of exercise to use there is an underlying principle that will test the appropriateness of any activity. It is: Does the exercise simulate the skill you are trying to teach. Ask yourself, "What does the student *do* in this exercise? Is it what I'm teaching?" For example, as discussed in Chapter 1, in "People Skills," students were asked to choose the response that paraphrased what the customer said:

Customer: As I was saying, my wife and I just moved into town. Got a new job, new car; ain't got no house yet. Been looking around. Kind of found a place. Thought you folks might be able to help us out.

Loan Officer: Response #1: Sounds like you want a home loan.
Response #2: How does your wife like it out here?

Let's examine carefully what is really happening in this example. Choice 1 does paraphrase the customer's response. However, choice 2 is so far removed from paraphrasing that the answer becomes obvious. Choice 2 is, in fact, a non-clarifying activity called "getting off the point." The student is choosing between a form of clarifying (paraphrasing) and a form of non-clarifying (getting off the point). A better exercise would test the difference between the two kinds of clarifying, paraphrasing and amplifying. For example:

Identify the response that best *amplifies* the customer's question:

Customer: As I was saying, my wife and I just moved into town. Got a new job, new car; ain't got no house yet. Been looking around. Kind of found a place. Thought you folks might be able to help us out.

Loan Officer: Response #1: Sounds like you want a home loan.
Response #2: Of course we can help, and we have three different home loan packages we can discuss.

In that example the student must distinguish between the two types of clarifying. Therefore, the exercise really is simulating the ability to recognize distinction between paraphrasing and amplifying.

As the writer of the original question, I cringe every time it comes on. In spite of the high probability that the wrong answer ("How does your wife like it out here?") would be given in the real world, in the context of the question it's obviously wrong. From this mistake, let's expand our simple test for evaluating exercise construction. After you create an exercise, ask yourself these questions:

1. Does it simulate the behavior I'm trying to teach (whole or in part)?

2. If it simulates *part* of the behavior, is it the right part?

3. Are there other exercises that simulate all the other parts so that the student learns the whole skill?

4. Does the exercise present enough choices to make it meaningful?

5. Is the exercise difficult enough to make the student consider the principles involved or is it obvious?

6. If you use humorous wrong answers, are there enough *other* wrong choices to teach the concept? Because humorous wrong answers are obvious, and obvious wrong answers don't teach.

Keeping these principles of evaluation in mind, let's proceed to the next important segment of an interactive video program—tests.

5

Test Construction and Evaluation

Unlike exercises, tests are designed not to teach, but to find out if the student has learned. For that reason there is no remediation following each question as there would be in an exercise; instead, questions are presented one after the other, the right and wrong answers are tabulated, feedback for all questions is presented and a score is given. Finally, remediation is provided if necessary. In this chapter, we will look at the various testing and options that can be integrated into an instructional program.

TEST FEATURES

Unit tests are the tests given at the end of each lesson. Final examinations, at least in our terminology, are tests on the entire interactive video program. The final exam is the last activity the student will complete before leaving the program to perform the new skills in the real world. Therefore, it should integrate all the steps being taught into the fullest possible simulation of the activity to be performed.

Another important aspect of a good final exam is that, like the lesson itself, it should be *criterion-referenced*. The key element in this technique is the criterion item, which lists not only what the student will do but how well. Ideally, all criteria for proper performance should be tested in the final exam. If a student is to perform a certain task and do it within a given time frame, that time frame should be included in the test. Many interactive systems have built-in timers which make this easy to do.

As noted earlier, the final examination should offer a true simulation of the behavior being taught. Interactive video simulation can never replace real-world performance, but it can often duplicate it very well. The challenge for a creative instructional designer is to construct tests that come as close as possible to actual performance. There are times, however, when the skills you are teaching may require live simulations in addition to those on the disc.

For example, "People Skills" taught principles of interpersonal relations and methods for applying them. In the final exam for "People Skills," students were asked to identify the best responses to comments from a customer during a simulated banking transaction. In addition to the video simulation, students were given a script which they could use to role-play the interaction with the customer. This added activity, in our opinion, greatly strengthened the students' ability to transfer the concepts to their job and offered a truer test of their ability to perform the skills we were teaching.

TEST STYLES

There are different types of tests that can be designed and integrated into an interactive video program. These can be classified according to test styles and test purposes. For the sake of simplicity, let's divide test styles into still frame tests and motion tests; we will divide test purposes into testing for understanding (concept tests) and testing for performance (simulation tests).

Still Frame Tests

Tests given at the end of each lesson can often be designed as single or still frames, since they are usually presented as some form of graphics and text. Seven to 10 still frame multiple-choice questions can be used to determine whether a student is ready to go on to the next subject. Figure 5.1 shows some sample test questions from the "Debits and Credits" and "People Skills" programs.

Figure 5.1: Examples of Multiple-Choice Single Frame Test Questions

```
"Content" is expressed by:
    1. Tone of voice
    2. Volume
    3. Clear details


All of the following phrases are examples
of stroking except:
    1. "That's a very good point."
    2. "I disagree."
    3. "You look very cheerful today."


To conclude the transaction you could:
    1. Disagree
    2. Change the subject
    3. Summarize
    4. Clarify
    5. Paraphrase
```

Figure 5.2: Example of Still Frame Visual Discrimination Question

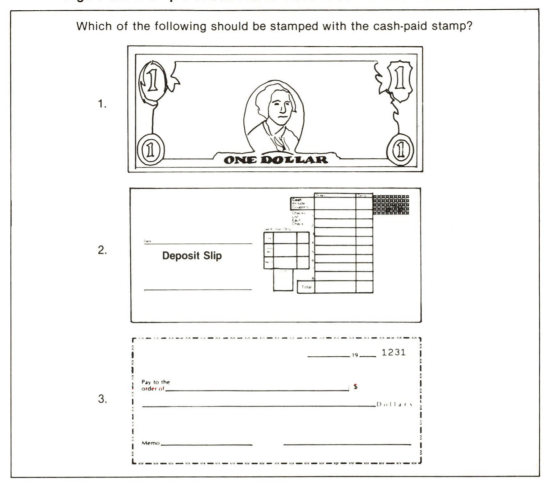

Still frame tests can also include questions on visual discriminations, such as the example given in Figure 5.2.

Answers to still frame tests can be presented simply by repeating the question frame with the correct answer highlighted or pointed out in some other way. Answer frames should be separated from the questions by some kind of introduction. This can either be given by the program narrator or shown on a still frame. Answer frames can be timed to stay on the screen long enough to be read through *slowly*, or a control instruction can be added to the frame; e.g., "To go on, press number 1."

Motion Tests

Using motion sequences in your test items allows you to represent the skill most realistically. When the skill itself requires motion, there is certainly no better way to represent the activity than to show it happening.

In the "Debits and Credits" program, we had to dramatize the inappropriate use of stamps by tellers who were leaving the "workstation." Since stamps are a "control item," they have to be locked up in the cash drawer when the teller leaves. Tellers very often walk away and leave their stamps on the counter where they can be stolen, or carry the stamps with them only to put them down in some inappropriate spot and lose them. None of this is easy to show in a still frame, but both the possible actions and their dangerous consequences can be well documented through the use of interactive motion sequences.

In final exams (especially where the highest level of simulation is required) motion sequences add enough visual "noise"—the distractions often present on the job— to provide a truly integrated test of the student's ability to perform the required skills.

TEST PURPOSES

Concept Tests

Concept tests are administered to determine whether students understand the principles involved in performing a skill. They do not test the actual performance itself. For example, let's say there are six steps involved in changing a spark plug. If you wanted to test for an understanding of this concept, you could have students do any of the following:

1. List the steps.

2. Identify those steps that are *part* of the process.

3. Identify those steps that are *not* part of the process.

4. Arrange the steps in the correct order.

5. While a spark plug change is in progress, identify procedures being performed correctly or incorrectly.

6. After a spark plug change is shown, list the procedures done incorrectly.

7. Answer specific questions about critical discriminations in the process; for example, which wires to attach to which plugs to assure proper firing order.

Note that at no time are you asking the students to change plugs; rather, you are testing their grasp of the concepts. With an understanding of the concepts involved, students will probably have an easier time performing the task in real life.

Simulation Tests

Simulation tests offer a visual enactment of the skill being taught and allow you to come very close to testing students' hands-on performance. For example, in the lesson on spark plug changing, you might move through the procedure to a decision point, then ask, "Now what do you do?" Students can then be given three choices:

1. Take out all the plugs.

2. Take the plugs out one at a time.

3. Take out one plug and check it.

Depending on the response given, the lesson would then proceed down one of three paths which eventually begin crossing, resulting in the classic spiderweb design discussed in Chapter 4. Deciding one of three courses of action is, in fact, the necessary behavior needed at that point. You could not, of course, simulate the actual screwing in of the plug which is a critical activity as well. Unfortunately, screwing in a plug is a tactile skill: you have to feel the plug screwing in smoothly. Things that require learning a certain feel are among the very few that cannot be simulated in video.

Video simulation tests are excellent ways to determine whether students have learned from the program. Since simulations can portray the consequences of correct and incorrect answers, they make excellent transfer exercises. Transfer exercises (as educational technologists know) are learning experiences designed to help students bring the skills they have learned back to their real world jobs.

When designing simulation tests, it is important to include the entire job with all its details in a realistic setting. The goal is to let students come as close as possible to real-world performance. The discussion under "Documentary Approaches" in Chapter 6 explores some video styles that lend themselves to simulation tests and transfer exercises.

TEST SCORING

There is a growing feeling that the developmental testing stage of interactive video —when the program itself is pretested with an audience—is far longer and far more demanding than was originally thought. Interactive video programs require extensive testing in final or at least check disc form before they are replicated in quantity. Nowhere will the effects of this pretesting be felt more strongly than in the area of test scoring.

In test scoring, the number of right answers or allowable misses required for passing is critical. Some skills require perfect or near perfect performance on the part of students. In other skill areas, however, since setting a passing percentage begins with an arbitrary decision, test scoring should be adjusted after the program has been reviewed extensively. Any adjustments that must be made may require a change in the disc control program or even in the video disc itself, but it is worth it. The alternative

Table 5.1: Guidelines for Test Scoring

Kind of Test	Test Style	Test Purpose	Typical Number of Questions	Typical Allowable Errors
Unit test	Still frame	Concept and simulation	7 to 10	1
Final exam	Still frame	Concept	12 to 20	3
	Motion	Simulation	Varies	3 to 5

may be an ineffective program or one that contains misleading questions that send students into endless remedial loops.

Test pass and fail standards will of course vary from program to program, but some rough guidelines are given in Table 5.1.

Testing the programmed check disc with the real-world audience for content and instructional effectiveness before quantity replication may seem like a time-consuming, expensive process. However, there is no other way to ensure the instructional effectiveness of an interactive video program.

TEST FAIL AND PASS MENUS

Once you have created a test that assesses students' comprehension of the material, you will want to determine where they will go if they have passed, or what kind of remediation they will get if they have failed. Here, as elsewhere, you can allow students to choose from a variety of options depending on their level of skill and rate of learning.

Test Fail Options

Basically, a unit test failure requires a review or repeat of the whole lesson. However, you may want to give students a choice of review options. Figure 5.3 presents a detailed schematic of a typical lesson. The branching for the exercise feedback is not shown, but you can see three demonstrations (DEMOS) broken up with

Figure 5.3: Schematic of a Typical Lesson

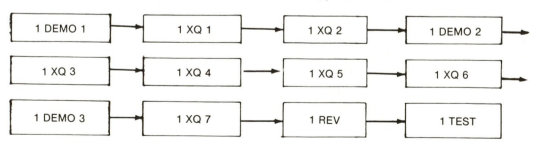

seven exercises (XQs) and a final review (REV). There are three logical ways to review this lesson:

1. Review the entire lesson (all demos and exercises);

2. Work through just the review and test; or

3. Just take the test.

I would bet that the last choice, the test-only option, would be the most popular with students; however, it is probably the least effective. After all, the only remediation that will have been provided is the answers to the test. This reduces the effectiveness of the program since, in essence, you would just be teaching for the test.

If students want a quick way out, they can take the second choice—the review and then the final test. Reviewing the whole lesson is a very admirable option, though probably not a very popular one. Figure 5.4 represents a compromise: The program is playing the demonstrations but skipping the exercises. It will play 1 DEMO 1 then automatically skip to 1 DEMO 2, to 1 DEMO 3 and then to 1 REV. What you have offered your students is a chance to listen to a lecture on the subject. This may be an excellent and popular remediation technique.

Another option is to offer a menu of the different subjects in the lesson. For example:

> You can select an individual subject which seems to be giving you trouble:
>
> Press #1 to review the concept of debits; press #2 to review all debit forms; press #3 to review the concept of credits; press #4 to review all credit forms; or press #5 to review the one exception to the debit/credit rule.

Finally, you can offer students a break frame, in which the machine pauses to give the students a chance to take a break. All of these options can be presented in a test fail menu. Figure 5.5 presents the script for such a menu. It shows both the text that would appear on the screen and the narration which would accompany it.

Figure 5.4: Demonstration and Review Option

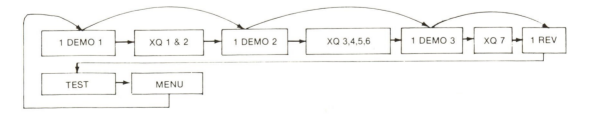

Figure 5.5: Test Fail Menu and Narration

Graphic	Narrator
You can review:	You should review the whole lesson; there are four ways to do that:
1. whole lesson	1. review the whole lesson
2. lesson without exercise	2. review the lesson without the practice exercises
3. key ideas	3. review the key ideas
4. individual concepts, or	4. review individual concepts which you think are giving you trouble, or
5. take a break	5. you may want to take a break
Indicate your choice.	Indicate your choice.

Test Pass Options

Test pass options are much simpler to design. Students can proceed to the next lesson, choose among several lessons or go right to the final exam. There is one creative alternative: to offer students a review of the key ideas (review sections) from all the lessons, before going on to the final exam. People who pass tests should also be given the option of taking a break. The script for a test pass menu incorporating all these choices is given in Figure 5.6.

Pretest and Preview Options

Instructional interactive video programs will usually be viewed by students of varying skill or knowledge levels. Therefore, it is often useful to provide a pretest to enable students who are already familiar with the material in one lesson to go on to others.

One form of pretest is to start each lesson with the option of taking the lesson test first. If the student knows the material, he or she will pass the test, and can skip the lesson. Another is to provide just the review and test for those trainees who want a slight brush-up. Finally, you can design the program to begin at the review, which serves as a *preview* of the lesson, followed by the entire lesson. As redundant as that seems, it does satisfy those who subscribe to the maxim "Tell them what you're going to tell them, tell them, then tell them what you told them."

FINAL EXAMINATION SCORING

There should be a tabulation mechanism for scoring the final exam, as well as all the other tests in the program. Unlike the unit test, the final exam is a true measure-

Figure 5.6: Test Pass Menu and Narration

Graphic	Narrator
Choose:	You have some choices here:
	You can take the lesson on balancing or
1. the balancing lesson	the lesson on stamps, if you haven't
	already done so. If you have, you can pro-
2. the stamps lesson	ceed to a review of the key ideas for the
	entire program before going to the final
3. review of all key ideas	exam.
4. the final exam	If you don't feel you need a review, you
	can go directly to the final. We'll even let
5. a break	you take a break before you do any of
	that.
Indicate your choice.	Indicate your choice.

ment device and, if true to its real purpose, should exist only to determine if the student has passed or failed the program.

Of course, the ability of the computer to do innumerable calculations allows the system to tell the students which questions they missed and what their percentage scores were. However, the general use of test data relating to *final* exams is to move students out of the classroom. Those who fail are supposed to take the program all over again.

Creative programmers may want to devise mechanisms to identify lessons or even particular concepts which are giving the student trouble. The students can then go back into the program and deal only with the sections that have prevented them from mastering the material.

Final examination data also help validate test items and testing techniques, and they should be studied carefully so that the program can be improved in the future.

PROGRAM EVALUATION

Evaluation is a way of answering all the questions that will be asked about the interactive program you have created: "Did anyone like it? Did they dislike it? Did anyone learn anything? Did it improve things on the job? How did it effect the bottom line? And, oh, by the way, *was it worth the money?*"

Developmental testing, discussed previously, is carried out after the check disc is made but before the program is in full distribution. It's a system of tests designed to work the bugs out of the system. It has nothing to do with measuring program effectiveness. Evaluation takes place when the program is in the field. There are several types of evaluations. Some are done as the program is being presented, others are done immediately after the program is over, others are completed back on the job weeks or months later. Ideally, the result of any evaluation is an honest report that shows column after column of positive data on improved performance.

Few systems of evaluation are completely reliable and many can be misleading. Still, it is possible to have some sense of the program's effectiveness, working with the levels of evaluation discussed below.

Levels of Evaluation

I have always identified five different levels of evaluation for instructional programs. All of them apply to instructional interactive video programs and many are appropriate for point-of-sale and game programs as well (see Chapter 8). Ranging from simple to complex, they are:

1. Do users like the program?

2. Can students pass a pre/post test on the skills taught?

3. Can they perform the skill immediately after training?

4. Can people perform the skill back on the job?

5. What is the effect of improved performance on the company's profits?

Level 1 evaluations are very basic, but they do give you feedback on how your users liked or disliked the program. This type of evaluation, of course, doesn't tell you how well the program worked or how much anyone learned.

Level 2 and level 3 evaluations may seem similar, but there is a big difference between them. Think of it in terms of a written driving test versus an actual driving test, or driving simulation.

Level 4 evaluation, whether people can perform the skill back on the job, is critical. Ideally, this should be measured through on-site observation by training analysts. This will require a manpower commitment but it provides a much more concrete measurement.

Many people feel that level 5 evaluation is difficult, if not impossible, to compute. The problem lies in being able to make a clear link between training and increased profits. For many years, a good training program's worth was said to be 10 times the cost of producing it. Thus, if a training program cost $100,000, the company would have to save or earn $1 million to cost-justify the program. All worth ratios in these cases must be based on a decision by the company to attribute some percentage of new profit to the success of the training. Unfortunately, the dollar value assigned to a training program is arbitrary.

A good argument can be made that *all* increased profits from a new product, for example, result from training. After all, untrained personnel may not be able to sell the product at all. Unfortunately, most companies will not accept that argument. They conclude, instead, that bottom line profit is just too complex a figure to tie directly to the cost of training. They recognize level 5 evaluation as an ideal, as an area so difficult to quantify that, in all likelihood, it will be quite misleading. If people need some yardstick or benchmark for value, simple metaphors seem to be very believable: for example, "If this training sells five new limousines, it will have paid for itself" is more believable than "There has been a 50% sales increase since the new model was introduced. We attribute 30% of that to advertising, 50% to the economy and 20% to training."

Tabulating Data for Evaluation

One of the great benefits of interactive video is that it has a built-in capability for data collection, tabulation and analysis—the computer. The natural ability of the computer to tabulate may be one of the best reasons for tying an external computer to an interactive video disc system.

The computer is best suited for level 2 and 3 evaluations. It can be used to tabulate the scores of the pretest and post-test, and compare the two. It can also score simulation activities, keep track of who passed or failed and of how well each student did. Further, the computer can be programmed to analyze data about those questions which were most often answered correctly or incorrectly and about questions that are problematic and need to be revised.

The computer can provide a printout of test results and correct answers. By accessing a data base that is specifically designed to offer feedback tailored to individual responses or response patterns, the computer can offer students printouts which provide additional clarification on individual answers, as well as an analysis of the student's performance.

You can design your own evaluation program by working closely with a computer programmer. Evaluation/software packages are also available from many of the companies that manufacture the interface systems that link the computer with the disc or tape player. Check with local computer hardware dealers or with manufacturers directly.

If you do not have an external computer, you will have to rely on traditional paper-based systems for data collection used for program evaluation. The best way to create a paper-based evaluation system for an interactive training program is to work with a training analyst or test-item writer. This person should go through your script and pull out individual tests and exercises. A well-written script will probably provide test item material on the video side of the page. It will be shown as descriptions of graphics frames used for testing.

By typing out individual pages of unit tests, unit exercises and final exams, your writer can create an answer sheet that students will complete while working through the program. If you correct the papers and tabulate right and wrong answers, you can analyze the consistency of answers. If your audience is a manageable size (under 100 people), you can tabulate the data by hand. If the audience is larger, you may need to enter the data into a computer after all.

The activity of completing an answer sheet while operating a video disc player may seem cumbersome, especially in individualized instruction. (Actually, it does work very well when you are presenting the material to a group in a classroom situation.) However, the results of your analysis should be well worth the effort.

6

Script Development: Treatments

As with most video presentations, interactive video programs are produced according to a script. A large part of script development is providing a description of the pictorial and narrative elements in the program. The way these elements are treated (program style) and the summary write-up of that style are both called treatments. In this chapter, we will examine various treatments of video material to see which best suit interactive programs.

Interactive video treatments can be divided into the same categories as those of linear video. I like to divide video treatments into two very broad categories: simple (live-on-tape) video programs and full production programs. These categories reflect production style, cost and length of production, rather than intended purpose. Much of this chapter deals with the interactive applications of full production video treatments, but first let's take a look at various live-on-tape styles in order to understand why they are less suited to interactive video programs.

SIMPLE TREATMENTS

Live-on-tape or simple video styles are designed to be shot in real time, with no post-production. Post-production is an essential part of creating interactive video programs. During post-production numerous shots, sequences and still frames are edited into easily accessible order. So the best we can say about simple video treatments is that they can form a small *part* of an interactive program.

Simple video treatments are also generally ill-suited to "how-to instruction," since good "how-to" requires a mixture of shots including graphics, text frames and extreme close-ups, as well as time compression and other techniques that are not easily accomplished in live-on-tape or real-time production.

The following are the most common live-on-tape treatments:

- controlled interviews

- walkarounds

- real-time lectures

- real-time speeches

- role-plays

- "live news"

- roundtable discussions

The controlled interview is usually conducted by a professional interviewer in real time. If the person being interviewed begins to ramble, the interviewer cuts him or her off. There's no editing process which would let you go back to select the best comments. What the participants say is what you get—and if the professional interviewer is good enough, what they say will be pretty close to what you want.

A walkaround is a commentary or testimonial usually shot in limbo. It features an extremely skilled spokesperson, walking from one display or "set up" to another (the "set up" can be as basic as brochures on a podium). Walkarounds are used frequently in commercials.

One way to increase impact (and probably cost) is to hire a well-known actor to do the walkaround. This can give you a high-impact message without increased production time and is sometimes called the "celebrity walkaround." Another variation is the "executive walkaround," in which a company president, for example, acts as spokesperson. The purpose of the executive walkaround is again to give the message more impact. Unfortunately, since few company presidents are professional actors, the net result is that the president often looks foolish, the person who suggested the presentation gets transferred and the video department's budget is cut by 50%!

The real-time lecture has great potential as an instructional technique and could be a valuable element in an interactive video program. Of course, its effectiveness depends solely on the quality of the lecturer. A good real-time lecture format is to set up a classroom in the studio, complete with students and whatever audiovisual equipment is required. You can then switch live, using two or three cameras, among the lecturer, the students and the AV demonstration. This normally generates a good deal of visual interest. In fact, this format may, with considerable elaboration, lend itself to interactive video. (But the lecturer had better be tops in the field, because the amount of post-production time needed to fix up a bad lecture would have allowed you to use a far more efficient format.)

Simple video role-plays sound easy but they seldom are. Basically, simple video requires simple sets (or no sets) and no editing. To shoot a role-play or dramatization that requires no editing, you must have a perfect performance. This is very difficult to achieve and probably calls for endless rehearsals and dozens of complete retakes. This type of treatment is a shortcut to the full production dramatization which will be discussed later. It certainly isn't appropriate for interactive video. It's a shortcut, and why use shortcuts when everything else involved in interactive video is so time-consuming and expensive?

Live news treatments and roundtable discussions are excellent simple video formats, but they don't really suit interactive instruction. It's difficult to conceive of a program that would build an interactive sequence from a roundtable discussion or a newsroom format.

The fact is that *all* simple video formats are shortcuts. They attempt to create polished programs on short timetables and with *no* post-production. What simple video treatments lack in production, they try to make up in talent and preplanning. Sometimes they succeed. In fact, simple linear video can be very valuable. At Bank of America, the use of simple video raised productivity from 30 programs a year to more than 100. It evened out the production cycle by giving the media department something to do in the inevitable downtime between larger productions. Most important, these kinds of programs gave many bank departments access to low cost and effective communication vehicles. But in spite of their appealing price tags and timetables, simple video styles have an extremely limited role to play in the creation of interactive video programs.

FULL PRODUCTION TREATMENTS

Simple video programs are derived from live broadcast television. Full production video programs, on the other hand, are derived from film production techniques, or what is called "film style" shooting. The difference is that in "film style," scenes are shot more in order of "production convenience" than in chronological order. Therefore, scenes can be shot over and over again until they are right. The final assembly of the program—the post-production—is time-consuming and extremely demanding.

Full production video treatments can be divided into five categories, which apply to interactive as well as linear programs: didactic, dramatic, documentary, graphics and a combination format. Table 6.1 lists each treatment along with a definition and statement of intended use.

Let's consider each of the five full production treatments in detail, paying special attention to their use in interactive video.

Table 6.1: Full Production Treatment Categories

Treatment	Definition	Use
Didactic	Features a teacher or spokesperson delivering a message or demonstrating a product or procedure.	To teach mechanical "how-to," conceptual theories, cognitive skills (is not especially good for interpersonal skills).
Dramatic	This is story-telling, acting out examples with fictional characters. It can include humor, pathos, melodrama, high drama and many other subcategories.	Best for teaching interpersonal or other people-related skills. Of course, it is also the heart of entertainment video.
Documentary	Features real people doing real things. Here too, there are many variations.	Often better for conveying attitudes than procedures. This treatment is extremely believable, but unpredictable when used with controversial subjects.
Graphics	A style in which titles, charts, graphs and schematic diagrams provide all the visuals. Unlike the other treatments, a pure graphics approach has its roots in slide presentations and photography. A subcategory is the symbolic or abstract approach.	This is basically an informational tool. It works best when presenting data or numerical procedures, and can be extremely effective in teaching the inner workings of things. It is also effective in allegories, fables and similies.
Combination	Can include all of the above, using whatever works best for a particular activity. But remember, some stylistic consistency is important.	Generally, this style is very appropriate for interactive video, which requires didactic explanations of system operations, as well as graphics sequences and menus.

DIDACTIC TREATMENTS

A didactic treatment is a video format that uses a teacher on screen, as a narrator, commentator, demonstrator or all of the above. Interactive video instruction is certainly a didactic system; its purpose is to teach. Moreover, interactive video requires a good deal of *internal* explanation or narration—what I call "housekeeping." As we will see below, the use of a didactic treatment facilitates this.

The format for a didactic treatment may seem relatively simple: just put the teacher on the screen. But it is not as simple as it seems; you must be ready to answer many questions. In what kind of setting do you put the teacher? Does the teacher participate in the program's demonstrations? What style of speech does the teacher use to give feedback or instruction?

To answer the last question first, I like a teacher who uses a casual style of speech, with informal phrases. An instructor shouldn't come on too seriously or not seriously enough. Some balance is needed. Here are a few examples of style points to keep in mind:

1. There is nothing wrong with using "we": "Now we'll consider the three basic kinds of customers."

2. Positive feedback can be very straightforward and encouraging: "Very good. Well done!" "Nice work. You passed."

3. Negative feedback has to be somewhat restrained: "You didn't make it." "You don't seem to get the idea."

At Bank of America we received complaints about narrators who appeared to be too judgmental in their facial expressions, in spite of the fairly mild narratives they delivered. This does not mean you can't say "no." "No" is straightforward, not insulting or too negative.

The Representative Narrator

As you can see, part of the decision you make in choosing a didactic treatment involves the kind of narration you will use, and the type of person who will deliver it. I suggest someone cheerful, straightforward, frank, intelligent and attractive. Frankly, questions of age, sex, etc., relate only to audience acceptance. The audience will usually believe and accept someone most like themselves. For example, at Bank of America, women between the ages of 25 and 35 represented the demographics of the bank's teller population. So the "People Skills" disc featured just that kind of narrator.

Another advantage of using a narrator who is representative of audience demographics is that he or she can participate in the demonstration segments of the lesson without any explanation being required. For example, in "Debits and Credits" the narrator appeared in a real, operating bank branch. At times she sat at a loan officer's desk, at times she was behind the teller counter, at times she stood in front of it. Very often she would talk directly to the camera from behind a teller station, then suddenly turn to face an approaching customer and begin demonstrating the principles she had just defined by functioning as a teller. Figure 6.1 shows this portion of the program script. As this example shows, a narrator who is demographically suited to the students can move in and out of the situations used to demonstrate instructional principles without any rationale being required.

Once a narrator is introduced and talks to the camera, he or she can act as teacher as well. Why pay for two actresses (a narrator and a teller) when you can have them both in one performer? If you *need* another screen presence, a non-speaking bit-player may do the job.

The "Detached" Narrator

An alternative didactic style is one in which the narrator/teacher doesn't do any

Figure 6.1: Portion of "Debits and Credits" Script

Video	Audio
Narrator, to camera from behind teller counter.	Narrator: How does this work in practice? Let's see.
Narrator turns from camera to face approaching customer. Camera pulls back.	Narrator (to customer): Good morning. How are you today?
Customer pushes check and deposit slip across counter.	Customer: Fine, thank you. I'd like to deposit this into my savings account.
Teller picks up the check, looks at name. Turns it over.	Narrator: Would you endorse it please, Mr. Ripley?

demonstrating. He or she is totally detached from the situation and, in fact, may not be in the same location in which the job is being done.

There are several reasons for detaching the narrator from the demonstration. The most common relates to subject matter: it may not be fitting for the narrator to participate in certain kinds of demonstrations. In "People Skills," for example, we were dealing with interpersonal relations. It seemed inappropriate for the narrator to participate in examples of interpersonal conflict. Therefore, our narrator never acted out examples of conflicts with customers.

A second reason for keeping the narrator out of the demonstration is because the demo parts of the program have already been produced. This is always the case when you are converting an existing linear video program to interactive. In such cases, since you still need a device to do housekeeping, to give feedback, etc., you can insert an on-camera narrator. That narrator can be in a classroom, in a TV control room, in a learning carrel. You can also present the narration as "voice over graphics" or, as commonly happens, just graphics.

Interactive Housekeeping

As noted earlier, the term housekeeping refers to all the procedural explanations that must take place in an interactive program. These explanations can be given verbally by a narrator on camera, presented on silent, still frame graphics or through a combination approach, narrated graphics. I recommend narrated presentations of housekeeping information. Not everyone likes to read, so by having graphics read to your audience you will increase the chances of getting your message across. Narrated presentations do use up disc real estate since they involve both audio and video tracks. However, they are worth it since they will make your program more effective.

Table 6.2 lists examples of housekeeping information that must be presented in an interactive video program, along with their definitions and frequency of use, i.e., whether or not the item must be repeated every time the appropriate situation occurs. I

Table 6.2: Housekeeping Information Items

Item	Description	Example	Frequency of use
Definition of interactive video	To explain what inter-active video is and how it works.	This is an interactive video disc program. It's different from other educational programs you may have seen because you direct it yourself. From time to time the program will stop and ask you questions about concepts and procedures. It will not continue until you answer using this con-troller.	At start of the first, second and third discs in your program series; then, as an option; finally, not all.
Bridges	To make the transition from a description of principles to a demonstration.	How does this work in practice? Let's see:	At least for the first transi-tion.
Introduction to menus	To present chapter or branching options to the student.	You have some choices here. You can: _____ Indicate your choice.	Use with critical menus. That is, menus that are extremely important to the student's progress or that have details that need ex-plaining.
Introduction to exercises	To present exercises.	Now let's see if you understand that con-cept. Press the key pad number that identifies the answer which you consider to be correct.	With first exercise in each lesson.
Introduction to reviews	To present review sequences.	Here's a quick review of the lesson on _____.	Before every review.
Introduction to exams	To present tests at the end of each unit.	The following is a short exam on_____. Follow the directions on each panel. At the end of the exam you will find out your score and be given directions on what to do next.	Before every exam.
Fail menus	To give test failure information and to list review options.	You didn't make it. You should review the lesson. There are several ways to do that. You can _____.	After failing exam.

(Continued on next page)

Table 6.2: Housekeeping Information Items (cont.)

Item	Description	Example	Frequency of use
Pass menus	To indicate passing a test as well as present options for further work.	You passed! If you haven't taken the lesson on _____ you'll want to do that now (or whatever options are available).	After passing the exam.
Introduction to final exam	To present the final exam and explain procedures.	The following sequence will show you several transactions. During the transactions you will be asked to make several decisions about the procedures being followed. Your ability to make the correct decisions will indicate your knowledge of the course material.	Before the final.

suggest you review this table during the creation of any didactic interactive video program. It is virtually impossible to over-explain, but it is very easy to forget to include all the appropriate instructions.

DRAMATIC TREATMENTS

Dramatic approaches are, essentially, story-telling, and as many of the world's greatest teachers have known, story-telling is a very good way to get your message across. Dramatic approaches are a valid way to teach, not only for training programs, but for the instructional elements of interactive point-of-sale and game programs as well. There are many different types of dramatic approaches, each with its advantages and disadvantages. There are also some special considerations to keep in mind when using dramatic video treatments as part of an interactive video program. We will examine some of the major dramatic styles with respect to these issues.

Realism

Realism is the easiest dramatic style to produce, since it is the least demanding. Its strength in terms of interactive video is that it presents an accurate demonstration of the skill or procedure being taught. The characters present models of correct behavior, the story gives a background or context that makes everything easier to understand and, taken as a whole, the lesson becomes an extended example of the principle being taught.

The problem, even with a realistic dramatic treatment, is that it can contain extraneous material that interferes with the content—the content gets lost or buried in an irrelevant storyline. While some extraneous storyline—or fleshing out of the plot—is necessary, writers and producers must resist the temptation to get sidetracked from the real purpose of the program. An interactive program should be entertaining, but not at the expense of content. Let me add that dramatic treatments of any type require very skilled actors, directors and writers. Work by a nonprofessional will result in a very unmotivating product.

The greatest strength of well-done dramatic treatments is that they do motivate viewers, encourage them to pay attention and make them care about the product, procedure or skill being presented. Realistic dramatizations are even more effective in this respect because, unlike other styles of presentation, you can have total control of the situation. In a realistic dramatization, you can get the exact words, actions, reactions, attitudes and behavior models you want, because you stage them.

For example, if you want to demonstrate a certain sales approach, the best way to achieve the exact effect you want is to script the approach word for word and have one actor deliver it to another. It is very unlikely that the same effect would be produced by following a real salesperson around with a camera. This especially applies to interactive video, with its varied outcomes. For example, you might need four of the most typical, but different, reactions from a person in a given situation. The best way to accomplish this is to script the situation, script the dialogue for the four responses, and have an actor present them.

Characters in realistic dramatizations can play varied roles in interactive video programs. You can have students evaluate the correctness of the character's behavior in a given situation. For example, a typical exercise might begin with these instructions:

> How would you evaluate the behavior of our teller in this situation? Please indicate 1 for very ineffective, 5 for very effective, or one of the ratings in between.

As noted, the student can also see the consequence of a given response in a given situation. Characters can also talk directly to the student and pose questions. For example:

> Customer (to camera): And what was your clue to *my* feelings?
>   1. tone of voice
>   2. volume, or
>   3. words that express feeling

Characters can also provide feedback. In the course of providing feedback, the character can even go so far as to comment on the story, the lesson or the whole learning situation. Involving characters in the interaction increases the interest and motivational impact of the training.

Humor

Take every negative associated with dramatic realism, multiply it by 100, then add one limited advantage and you're dealing with humor. Humor is the most dangerous form of communication there is, essentially because it is so subjective—one person's joke is another's insult. When you are training people within an institutional setting, you must be careful not to insult them or they will take a dislike to you and the program and, consequently, won't learn. There is another risk to using humor. Even if you are able to come up with a series of humorous situations that don't demean any group, person or class of people, you may still convey a feeling that the skill, job or procedure is undignified or not to be taken seriously.

In "People Skills" we used a humorous samurai warrior who presents his sword at the counter when the teller asks for ID impolitely. The scene did not evoke any controversy, but it could have.

There are some benefits to using humor properly. People do like to laugh and judicious use of humor can make a program more interesting. Also, correctly used, humor can act as a tremendous memory aid. "Sesame Street" has had phenomenal success using humor in an instructional setting.

One use of humor specific to interactive video is the humorous multiple-choice question. Usually, the humorous answer is presented first in a series of answer choices, several of which are closer to the correct answer. If the choices are too obviously wrong, you will be criticized for not really teaching because the exercises are too easy. The following humorous multiple-choice question was used in "People Skills":

> Nutty Professor (making a suggestion to a teller):
> Instead of this single customer line denoted by the red carpet I do believe and I further indicate . . . my proposal is that you should always allow individuals to choose whichever of the several transactional orifices which they deem to be in the most appropriate propinquity to their current logistical grounding as well as proceeding with the greatest alacrity.
>
> Teller (Choose the best response):
>   1. Huh?
>   2. You mean, you'd like to do away with our red carpet service?
>   3. Never mind the lines, what can I do for you?
>   4. Would you like to talk to our manager about it?

Fantasy

Fantasies are extended stories told about the content but at times and places unrelated to the location or time in which the skill will actually be performed. They contain all the advantages and disadvantages of other dramatic styles in addition to a few of their own.

The main advantage of a fantasy is that it has a timeless quality: It won't be dated by style, costume, attitude, etc. There was a time in the late 1960s and early 1970s when we had to remake shows every three years because, even if the content didn't become outdated, the clothing styles did. Set your lesson in the Old West, in the future on a space station, in the days of cavemen or some such time and you eliminate the need to update styles. Programs that might otherwise have a shelf-life of three years can then last indefinitely. Fantasy works especially well with timeless content, e.g., how to sell, how to treat customers, etc.

What's wrong with fantasies? You'll know the first time you present your sales training western and the executive vice president says, "But what's that got to do with *selling*? And, by the way, what did it cost?" Fantasies usually cost more than other kinds of programs because of the need for costumes, props and sets. Add cost to the fact that fantasies often seem irrelevant and you're in trouble. The argument that the show will be timeless may not offset the added $50,000 expense.

Many executives feel that the fantasy storyline detracts or distracts viewers from learning. The determining factor, of course, is how well the fantasy is written.

I wrote a fantasy for Walt Disney's educational division that taught the basic principles of economics when Huey, Dewey and Louie needed money to build their own playhouse. In order to obtain funds, they started a business manufacturing doorstops, the most needed commodity in Duckburg, a city built on a hill where doors kept slamming in people's faces. The situation set up the presentation of all content: 1. serving customer needs; 2. supply and demand; 3. the factors of production; 4. marketing; and 5. economic cycles.

A story *must* set up and provide a relevant context to the educational content, or the advantages of building interest and being timeless are lost. The essence of the message has to be at the heart of the dramatic situation. If you choose fantasy, the fantasy setting has to facilitate learning, not hamper it.

Fantasies work well in interactive programming, because interactive video usually mixes at least two styles—one of which is didactic. One of the other styles can be fantasy, the exaggeration of the consequence to the extreme. The "larger than life" aspect of fantasies can be used in multiple-choice responses to dramatize the outcome of a particularly bad or good choice. It should be noted, however, that if fantasy feedback is used it should be used with some regularity and, of course, it must be done well.

In point-of-sale, fantasy settings for product displays are quite useful. "Share the fantasy" is a successful slogan of one ad campaign. It seems very logical to allow customers to branch to a product display showing one or more products in an adult fantasy setting: for example, the Caribbean cruise, the dream home, the medieval banquet. If you must steer customers through pages and pages of products, consider offering the product fantasy option. Your customers will be grateful.

One weakness of fantasy as it relates to interactive video is that the expanded content needed to produce it—e.g., the variety of additional sequences that have to be produced to complete multiple choice exercises—may make the already high costs of fantasy climb even higher. Also, the fantasy theme may have to be carried into the still frame sections, increasing the cost of graphic sequences.

Table 6.3 summarizes the advantages and disadvantages of the use of realistic, humorous and fantasy dramatic treatments.

DOCUMENTARY TREATMENTS

A documentary is an attempt to capture the reality of a situation by showing real people doing real things. In the mid 1960s it was fashionable to begin any discussion of documentaries with the question, "Yes, but what do you really mean by the term documentary?" The origin of the question was the debate over how much a film editor, narrator, director or TV producer could color reality by manipulating content.

Editing, cinematography and directing techniques *can* control the reality presented in a documentary, but just as often the reality leads a life of its own that no one can control. This is the problem when the documentary style is used in interactive video—it is difficult to control, especially when varied outcomes are needed. As noted earlier, the best way to teach selling techniques is *not* to follow an expert salesperson around with a camera. You never know if you'll get the kind of presentation you really want, in the way you really want it.

The same applies to greeting customers, changing spark plugs or any other skill. Experts operate by their own rules—they take shortcuts and often work by instinct, something trainees can't do and shouldn't attempt. People starting out must simplify the task, take extra steps and look for more details, not less.

In addition to the unorthodox techniques of experts, documentary production brings producers face to face with the concept of "noise." In learning theory, noise is what gets in the way of a trainee's receiving the necessary stimulus, i.e., distractions. When you are trying to depict a detailed procedure realistically, much of that reality is noise. Many details in the environment are not specific to the procedure, and they make the procedure hard to focus on.

To teach, you must begin by presenting the procedure without the noise. That way, the students get to see exactly what you want them to do. Then, gradually you can add noise—the details of reality—so the skill will be seen in a context and will be easier to transfer to the real world.

Table 6.3: Dramatic Treatments: Realism, Humor, Fantasy

Advantages	Disadvantages
Realism	
Provides realistic demonstration	Can contain:
Presents models of correct behavior	Irrelevant or extraneous material
Motivating	Controversial material
Highly controlled	Unprofessional acting, dialogue, direction
Accurate	
Provides context and background	
Humor	
Acts as mediator or memory aid	Can contain:
Enjoyable, encourages learning	Irrelevant or extraneous material
Motivating	Offensive material
Provides context and background	Controversial material
	Unprofessional acting, dialogue, direction
	Can reflect or convey lack of seriousness in the subject
Fantasy	
Timeless (avoids changes in style)	Can contain:
Acts as mediator or memory aid	Irrelevant, distracting content
Enjoyable, encourages learning	Controversial material
Motivating	Unprofessional acting, dialogue, direting
Provides context and background	Expensive to produce
	Can complicate all aspects of interactive production
	Can convey lack of seriousness in the subject

The instructional process I have just described is called shaping and can be outlined as follows. To teach a complex skill:

1. Present the procedures in the abstract without any distracting irrelevant details (noise).

2. Present several exercises which gradually add more and more distracting details.

3. Test the procedures by presenting them in the context of reality (all the distractions of the real world).

The documentary approach would fit into step 3 of that shaping process, where you show reality. Thus, an interactive video program might begin with a diagram of a procedure. In the next step, the procedure would be acted out using actors in a limbo (or nonspecific) setting. The final step in the program would test students' grasp of the procedure by presenting it in the context of reality.

The noise that makes the documentary seem so very real gives it an additional extremely valuable property. Because documentaries capture irrelevant details that are present in reality, they are believable. The details are cues picked up by our subconscious that tell us *this is real*. Documentaries are the most believable of all styles. While following a sales rep around with a camera may not be the best way to teach *how* to sell, a series of statements by real customers on what they want sales reps to do will help students want to sell correctly.

Documentary approaches should be viewed as a valuable component in complex interactive video programs. However, an interactive program done entirely in a documentary style probably would not work.

Typical uses of documentary techniques in interactive videos would include: motivational overview sequences which provide background and introductions; expert commentaries offered in the context of the entire program; expert commentaries offered as optional asides; or documentary-style presentations of policies or procedures used as part of a final exam. The following are examples of each technique.

Motivational Overviews

This could be a montage of customer, ex-customer and noncustomer statements telling exactly how they want to be treated by salespeople. When interspersed with real-world footage of day-to-day operations, this should provide good motivation for learning how to sell. Likewise, a one-minute commercial about what a company is, how it does business, etc., is a good addition to a point-of-sale program. Just remember that these motivational overviews don't teach; the best they can do is help the students *want* to learn.

Commentaries in Context

This is a way to provide the high credibility of an expert commentary during your program. We used a psychologist in "People Skills" to explain the three major types of customers. The psychologist's commentary was a documentary. Scenes of people exhibiting the behavior he described could also be considered part of the documentary style. Again, remember that this activity by itself is not enough to demonstrate the principles let alone teach them. It only provides part of the demonstration and requires other, more complete support.

Commentaries/Examples as Asides

These are expert commentaries set off from the mainstream of the program. For example, as the result of a multiple-choice question, students can access an actual expert in the following manner:

> Would you like to hear exactly how a trained psychologist would explain the behavior we just witnessed? Press 1 for "yes" or 2 for "no." If you press 2, you will proceed to the next part of the lesson.

Documentaries of Procedures as Final Exams

In a final exam, a documentary sequence can show a procedure in a real-world context. The sequence is only interrupted by questions designed to find out if the student can identify the correct procedures. For example:

> Narrator (introducing the final exam):
>
> The following sequence will show you several transactions. During the transactions you will be asked to make several decisions about the procedures which the teller is following. Your ability to make the correct decision will indicate your understanding of the procedures presented in this program.

Documentaries Featuring Chief Executives

The chief executive officer of your company can be presented in a documentary style in your program. This use is not the corporate stamp of approval or the speech read from a teleprompter. Those styles are didactic: the boss is telling the audience something directly.

The set-up for a documentary featuring executives is an off-camera interviewer. The boss's statements are opinions which the viewers feel they are overhearing. The comments are heard by the audience but do not seem aimed directly at them. The net result is those little speech imperfections, pauses and details that tell the audience *this is real.* Documentary statements made by executives are usually far more believable and effective than speeches, since the boss is speaking in his or her own words.

An interesting variation on this style is the interactive interview. In the interactive interview a standard interview is conducted. Then, the interviewer is cut out of the interview and replaced by a question menu; i.e., a menu that lists all the questions stated in three or four words each. The computer program is set-up to take the viewer from the menu question to the appropriate answer, and back to the menu again. The result is that the viewer only asks questions that he or she is interested in. Thus, the viewer has more control and pays more attention. Table 6.4 summarizes the advantages and disadvantages of the documentary approach.

Table 6.4: Documentary Treatments

Advantages	Disadvantages
Extremely believable	Difficult to control
Motivating	Can lack focus
Expresses real opinions of customers	Too noisy for early stages of training
Provides high degree of realism since contains real-world noise	Can be confusing because it contains shortcuts by experts
Is an excellent transfer exercise	

GRAPHICS TREATMENTS

Programs that are primarily graphic are an outgrowth of the slide genre. The presentations can range anywhere from simple titles to full animation.

The advantage of a graphics approach is that it can simplify a process or activity to a remarkable degree. This makes graphics approaches excellent demonstration tools. They also eliminate the narrator's on-camera presence completely. This is especially beneficial for programs that will be translated into other languages or that attempt to take advantage of the multiple soundtrack capabilities offered by the video disc.

The large number of tests and exercises in interactive video programs mean that most interactive programs have a large graphics component. Be sure to allocate enough time and dollars for the proper creation of these elements.

I would like to encourage all producers who must deal with an abstract concept or a complex mechanical activity to consider using title graphics, illustrations or animation to give a clear demonstration of the concept or activity before adding the noise of the real world. Then, using the shaping process described previously, the program can gradually move toward a more realistic presentation of the information including all the noise present on the job. The inner workings of engines, various kinds of surgery, flow of paperwork or forms are just a few examples of the kind of subjects that work well using the graphics approach.

The major disadvantages of a pure graphics treatment occur in those programs that rely too heavily on this approach. Pure graphics styles, without the noisy realism of live footage, seem too abstract and students seldom make the transfer back to on-the-job performance. Nowhere in life do we really encounter cut-away schematics or animated flowcharts.

The danger of the graphics approach is compounded in interactive video production. There are, after all, 50,000 frames on a laser disc. Why not fill up every one of those 50,000 frames as still frames? The reason is that it is very difficult to teach motion without producing motion. Similarly, you have a better chance of selling things that move by showing them moving.

Video discs can give you thousands of still frames and graphics, an encyclopedic number of pages. But knock that laser out of that single groove and let it move past those grooves at 30 grooves per second, and you'll give your program something more than storage capacity: you'll give it life!

Table 6.5 summarizes the advantages and disadvantages of graphics treatments.

Table 6.5: Graphics Treatments

Advantages	Disadvantages
Excellent demonstration tool	Too simple
Simplifies complexities	Too abstract
Good at the start of noise-free lessons for beginners	Too dull and lifeless
Good way to create multilingual programs	

COMBINATION APPROACHES

Obviously, each video treatment style has its own strengths and weaknesses. As the preceding discussions suggest, a complete interactive video program is a combination of many forms. Table 6.6, which follows on page 88, shows an outline of a typical interactive program that combines several styles, indicating the one that is most suited to each activity.

Keeping these various video approaches in mind, we turn next to the scriptwriting process itself.

Table 6.6: The Combination Approach

Topic	Treatment
Opening menus	Graphics
Welcome	Didactic
What is interactive	Didactic
Rules of the program	Didactic
Why use the program	Documentary
Demo on task 1 in lesson	Graphics
Exercises 1 + 2 (on task 1)	Graphics
Demo 2 on task 2 in lesson	Dramatic
Exercise 3 + 4 on tasks 1 and 2	Graphics
Demo 3 on task 3 in lesson	Dramatic
Exercise 5 on tasks 1, 2 and 3	Graphics
Review	Didactic
Rules for test	Didactic
Test on tasks 1 through 3	Graphics
Test menus	Graphics/didactic
Rules for final exam	Didactic
Final exam	Documentary
Test reporting	Graphics
Epilogue/farewell	Dramatic

7

Script Development: Writing It Down

In interactive video, a large part of script development is concerned with accurate description of all the still frames, housekeeping segments and other elements that are unique to the medium. These unique elements mean that the development of an interactive video script calls for a number of skills not normally needed for traditional video scriptwriting. This chapter will review the mechanics of writing and organizing an interactive video script.

In my experience the tricky part of script development is not writing the script; it's getting the script approved. Let's face it, the people paying the bills get to tinker with the script whether they know what they are doing or not. With this in mind, the best course of action is to adopt a writing system that simplifies decision making, allows the client to focus on the important points and moves everything along more quickly. This "systems approach" is even more important when dealing with interactive video, where decisions are even more complex than in linear programming.

STORYBOARDS

Since many industrial/training video programs are based on very straightforward treatments, storyboards end up as illustrations of talking heads or people in very traditional settings. Storyboarding such simple concepts is really a waste of money. This does not mean that storyboards have no place in industrial script development. They do have a place—not to detail a story, but to sell a concept.

For example, say that you have an expensive show with a complex concept, and a client who is unable to visualize things. In this case, to get your idea across, you can produce *concept boards*, which dramatize the style of the program in six to 10 large illustrations. They give the setting and the treatment and, best of all, they give you a presentation for the client. If you can put on a clear presentation, you'll probably get the client to buy your idea. In my six years as head of Bank of America's media center, I sold every single concept I presented with concept boards.

Photographic storyboards are another effective technique. They are also developed at the treatment stage, but are created with 35mm prints that the writer or director can take and pin up on a board. Add a few magazine cutouts to cover areas not readily available, and you have quite a sales presentation.

In either case, the purpose is to sell your idea. The storyboards are used at the concept phase, and once they do their job, you can *then* begin detailed scriptwriting.

NARRATION-ONLY SCRIPT

This is another technique that helps get scripts approved in the concept stage. Use the traditional two-column technical/instructional script format illustrated in Figure 7.1, but only complete the audio part of the page. Provide a word or two to indicate any necessary visual description, but otherwise leave the video description completely blank. This technique will force reviewers to concentrate on the narrative, leaving the writer and director free to develop the visuals as they see fit.

Figure 7.1: The Technical/Industrial Script Format

Video	Audio
Narrator	**Narrator:** I'd like to explain an effective way to get your clients to approve your script.
	Remember the "Golden Rule" of video writing?
	It goes like this: "The first draft is never approved, so never put your best ideas in the first draft."

As unusual as this technique sounds, it is not at all unfair to your clients. They will have agreed to the visual treatment via an earlier concept board presentation. Approval of the narrative, then, is their main concern—one which you will keep in focus by the way you present the script.

If the client insists, you can follow the concept board/narrative script phases with a completed script. However, if you have established enough credibility, your client may accept the argument that detailed scripts are designed for technicians and are difficult for the uninitiated to follow.

THE INTERACTIVE SCRIPTWRITING PROCESS

Having reviewed techniques on storyboard development and script formats, let's focus on the special requirements of interactive video instruction. Development of an interactive video script is a matter of working through the design document and writing dialogue and visualization that match the activities the design calls for.

First, write the introduction, which must include an explanation of the interactive program and how it works, as well as a motivational overview (why the student should want to learn).

When you write the first demonstration, work on it until you feel you've presented a digestible amount of data (i.e., the amount viewers can assimilate without a review). Try to conform to the design document, but if you sense that the document calls for too much or too little, don't hesitate to include a bit less or more information. This may affect the overall design, so keep an eye on the consequences of your changes.

Next, proceed with examples, exercises and feedback segments. Review Chapter 4, which describes the techniques for creating instructional exercises. Make sure that you present at least one exercise on each individual concept. (You can teach two concepts with the same exercise. For example, if you ask someone to identify a certain form as a debit or a credit, you will be teaching both concepts: what is a debit and what is a credit.) Make sure, also, that you follow all exercises that teach *parts* of a skill with a single exercise that integrates all components.

The following are a few additional pointers to keep in mind as you write the exercises and tests for an instructional interactive video. (Again, we'll rely on advice from the Harless Performance Problem-Solving Workshops.)

- Present the behaviors determined in the learning problem analysis, and present them at the *right* level of simulation.

- In designing practice exercises, present cues that help the person learn.

- In criterion tests there should never be cues.

- Don't include trivial questions.

- You should ask for behaviors in the sequence in which they are performed on the job.

- You should not use fill-in-the-blank items unless the real world behavior is fill-in-the-blanks.

- Build your lessons with isolated practices that include more and more reality.

- Make sure that the final test integrates everything taught into a detailed simulation of the real job.

A Word About Quality

The heart of a great script is the quality of the writing. But interactive scripts require many different kinds of writing. As a result you, as an interactive writer, must be very versatile. You must also pay great attention to the quality of your writing, both in terms of clarity and creativity. Don't crank out your script. Rewrite it again and again, then ask yourself: What was I trying to say? Did I say it? Could I have said it better? Then, rewrite everything one more time.

Writing Dialogue

Interactive video is also an audio medium, so it requires spoken words and exchanges between people—all of which must be believable. Writing dialogue is an art form. Bad dialogue is not only obvious, but is usually the object of ridicule: it will destroy the credibility of your program.

There are many techniques for writing good dialogue. My favorites are good listening and good rewriting. If you listen carefully to the way people talk, to the words they use, to their speech patterns, you will build up a collection of words and phrases. If you have trouble keeping these phrases in mind, start a card catalog. Write down appropriate words and phrases so that you will have examples when you need them, instead of having to rely on memory.

When you are writing the narrative portions of scripts, be especially careful to follow the logical flow of the narrative when branching is taking place. Read through each branch independently to make sure that there are smooth transitions from one statement to the next.

My remarks about the logical flow apply to dialogue as well as to narrative. A person's response must match the statement to which he or she is responding, not just in context but in the choice of words. Here again, read through each branch with an eye and ear for smooth dialogue.

Writing Exercises

Chapter 4 discussed good exercise writing in detail. As a final note, I suggest that you be open and creative when writing exercises. Even though you will be following a rigorous set of rules, realize that creativity can shape those rules and demands into something very exciting.

The instructional design will determine the kinds of exercises chosen for a program. But the quality of each exercise is really the job of the writer. Of course, the designer has final approval and will probably have to be consulted on any changes that are to be made. However, you, as the writer, understand the flow of information, the continuity of the storyline, character development, etc. So you are in the best position to make detailed decisions about the copy.

THE FINAL SCRIPT: CONTENT IDENTIFICATION

The final typed version of the script has to include information that script writers have never had to provide before. Because the interactive system is able to seek out and play discrete segments of video from a definite beginning to a definite end, all segments must be labeled and their starts and stops must be identified accurately—to the *exact* frame.

The latter can't be done until the tape is edited by noting frame numbers for the first and last frame of each segment. But even in your early drafts it might be wise to provide a format where these frames can be noted. Figure 7.2 shows a sample final script format indicating the places where you should note frame number start/stop locations.

Figure 7.2: Final Script Format

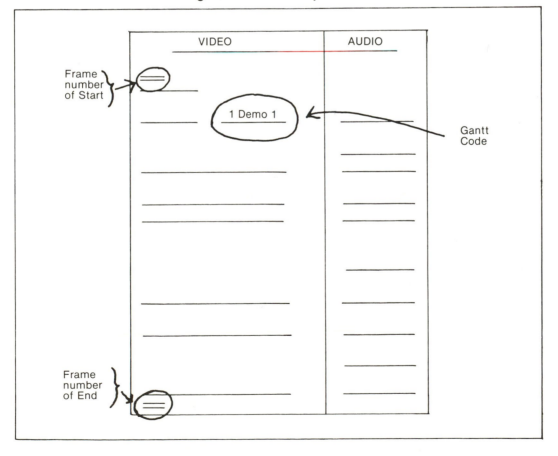

As noted throughout this book, it is critical to test, review and retest all materials. The script should be seen not only by subject matter experts but by people from the target population who can tell you whether or not the exercises work. With this type of input, you will be able to go into production with some confidence that your program will succeed.

8

Planning, Production and Post-Production

The actual production phase of interactive video is very much like standard video production. The interactive video director does not have to bring a vast array of new skills to the sound stage. What is required instead is greater emphasis on a few critical activities.

PREPRODUCTION

Preproduction, as we know, refers to casting, location and prop selection, costuming, equipment rental, etc. In interactive video, the preproduction activity that requires special emphasis is the extra effort the director must make to understand the script, its organization and its continuity. The director must trace every path through every possible branch to make sure continuity is always present.

Some people also recommend that directors study the final flowcharts of the program. The director must take every step possible to ensure that the script is understood. The best way to do that, of course, is to insist on a tight, clear, fully detailed script.

Casting

Actors always seem to enjoy doing interactive video programs. The variety of solutions to every situation requires actors to draw upon the full range of their acting skills, especially when doing interpersonal relations training.

Be sure that only the best actors are selected for critical roles. You could say that this is true in all production, but linear video lets you cheat. If you typecast, for example, you can get an actor who automatically does what his type does in every situation. However, interactive video is based on the premise that the character will do not only what is most likely in every situation, but will also be able to respond in several very different ways. To do that, actors must be more versatile.

Directing

Directors must have creative freedom to work. If the scripted responses don't seem natural, the directors and actors should be able to improvise. This may lead to making flowchart changes, but it will be worth it. Every step toward limitation is a step away from creativity and from a more natural, believable program.

During the shooting, it may be necessary for the director to explain the concept of interactive video over and over again so that the actors can understand what's going on. The following is a brief summary of what the director might say:

> This is a technique in which the audience can choose many out-
> comes for each situation, so we have to shoot separate outcome
> scenes for each possibility. The show won't even make sense
> when it's edited together. The computer program that lets the
> audience make choices is the ingredient that makes it all work
> correctly.

Continuity

Continuity is always a problem for programs shot film-style (out of order). The large number of components, the number of extra scenes, bridges, feedback segments, etc., make good continuity a critical issue in producing interactive video programs. You must have a person follow the script, making sure each scene *is* shot. Having snapshots on hand to remind yourself of costuming, props and actor positioning is also helpful. In planning for an interactive video production, budget for and hire a continuity person, even if you have never used one before.

Production Quality

The whole structure of corporate video seems to be set up to discourage the highest quality. Corporations seldom have seasoned video professionals on staff. Often corporate media personnel are on the way up, still learning. Corporations also tend to discourage risk-taking or anything that hints of extravagance or a waste of money. So to do something new, exciting or different can be quite difficult.

Of course, if you look at human history in general, the items I just mentioned may not be as critical as they seem. Certainly, masterpieces have been produced by unknowns, and great works are not always risky or extravagant. What is far more important is a commitment to quality by workers on every level.

If you are the producer, seek out workers who are challenged and motivated by the opportunity to deliver quality, and who will be as demanding on themselves as you will be. Don't allow shortcuts, and do everything you can to assure enough time for proper execution of each step in the production. Budget adequately so you have the resources to do things well. And above all, let everyone know that you care about quality.

POST-PRODUCTION: INTERACTIVE EDITING

One of the greatest discoveries made by students of film or video production is the tremendous creative rewards that come from the process of editing. More than a simple assembly of shots, editing makes a statement all its own about the content. It's not just what is in a shot that counts—there is meaning that comes from the way shots are put together.

The great Russian filmmaker Sergei Eisenstein was the first to identify the power of "putting together." He called it "montage" and said that it added a whole new dimension to his work. The formula "1 + 1 = 3" was one of the ways he expressed it: the third ingredient was added by the "putting together." Certainly anyone serious about film and television editing should become familiar with Eisentein's ideas about the power of montage: pacing, juxtaposition, shot selection, etc. Eisenstein's books on the subject, *Film Form* and *Film Sense*, offer tremendous insights for video editors (see the bibliography at the end of this book).

Interactive editing adds two new elements to the art of editing, and I think they are important enough to be presented with full appreciation of the already overwhelming power of editing as a production tool. These new elements are *user choice edits* and *redundancy*. Let's look at each.

User Choice Edits

An edit, of course, is when you put two separate shots together. As Eisenstein said, the way the two shots fit together adds something more than the content of each individual shot. If, for example, you show a shot of workers pouring into a factory (as Charlie Chaplin did in "Modern Times"), then you cut to a shot of sheep being driven to the stock yards, the two individual shots have added meaning that wasn't present in either shot separately: 1 + 1 = 3.

With interactive video, the sequence of shots can be chosen by the viewer. If you pick one action, the computer selects one of several different reactions and one of several different shots. That means that the viewer has something to say about the addend in the equation 1 + 1, and that choice could well lead to a very unexpected "3."

The situation takes on added complexity when the writer, in a moment of creative genius, notes that one of those responses is exactly the same as a response to another question. What you have then is the start of a spiderweb and of a problem. If the director studies the flowcharts, he or she may notice that the response to one question must also be used to respond to another. A scene from one setup must intercut with another. If not done carefully, there will be a jump between cuts resulting the editor's classic dilemma: the jump cut. And don't make the mistake of thinking that the subliminal search time will cover the jump. A jump cut is still jarring, even with a quick search sandwiched in. So, watch your cuts, especially if you do spiderwebs.

The task of avoiding jump cuts falls very heavily on editors of interactive video programs. In a video sequence showing an exercise question (XQ) and three feedback response frames (1XCF, 1XFF1, 1XFF2), the three feedback sequences will be placed in the order shown in Figure 8.1. (Note that only the first false feedback (1XFF1) follows the question physically. This is discussed later on.)

Figure 8.1: Video Sequence Edit

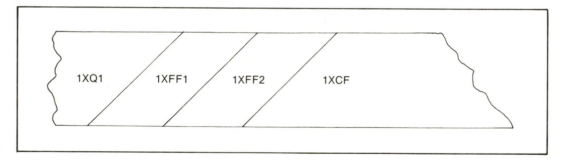

The editor must be sure that the editorial continuity is as strong when going from XQ1 to 1XFF2 or 1XCF, as in the case of 1XFF1. Yet the editor will never see the continuity, because it will only appear when the computer program is placed into the finished program and the program skips from 1XQ1 to 1XFF2. Therefore, user choice computer edits have to be worked out by the editor as carefully as physically contiguous edits. When the computer branches from one scene to another, it's an edit, and all the rules of good editing apply. If they don't, 1 + 1 won't equal 3, and the message generated by juxtaposing the two shots won't come across as clearly.

Redundancy

Redundancy is a useful technique for maintaining continuity. For example, say I present an interactive scene in which a bank customer loses his temper and a teller responds in two different ways. A still frame question appears asking the viewer to select the teller's best response, and the viewer probably takes some time to pick the best choice. In the meantime, any sense of the continuity of the dialogue may be lost. Figure 8.2 shows the script for this scene.

Figure 8.2: Script for Dialogue Followed by Freeze Frame

Video	Audio
Close up: angry customer	CUSTOMER (angrily): I demand that you start following the procedures to close my accounts.
Medium shot: teller politely Choice 1.	TELLER (sympathetically): You want me to begin closing all your accounts, is that right, Mr. Johnson?
Medium shot: teller angrily Choice 2.	TELLER: Let me get my supervisor. Maybe *she* can find a a way to help you.
Freeze frame. Two-way split of angry and polite tellers. Key graphic of choices: 1. You want me to close your accounts 2. Let me get my supervisor	NARRATOR (VO): Choose the most effective response.

How does an editor reestablish the continuity lost at the freeze frame? I suggest that when the student chooses a response, repeat the scenes leading up to the outcome. For example, XCF (the correct answer and feedback sequence) becomes the entire sequence shown in Figure 8.3. Thus, the correct response not only shows the outcome of the exchange but repeats the exchange itself.

The redundancy of repeating the entire exchange may use up valuable disc space, but it more than makes up for this by clarifying the conversation and making sure none of the continuity is lost. Of course, redundant editing is not always necessary.

Figure 8.3: Reestablishing Continuity

Video	Audio
Close up: angry customer	CUSTOMER (angrily): I demand that you begin following the procedures to close all my accounts.
Medium shot: teller, politely	TELLER (sympathetically): You want me to begin closing all your accounts, is that right, Mr. Johnson?
Medium shot: customer more softly	CUSTOMER: That's what I said, didn't I?
Medium shot: teller, obediently	TELLER: Yes, Mr. Johnson.
Close up: customer, far less angrily	CUSTOMER: But all I really want is to be able to deposit these funds today.

When the continuity is clear, or the exchange is long and painful, direct feedback without redundancy may be best. For example, after a rather lengthy statement by the "nutty professor" in "People Skills" (the example appears in its entirety in Chapter 6), the viewer is given these choices for the teller's response:

> Teller (Choose the best response):
> 1. Huh?
> 2. You mean, you'd like to do away with our red carpet service?
> 3. Never mind the lines, what can I do for you?
> 4. Would you like to talk to our manager about it?

Viewers who select choice number 2 branch directly to the correct feedback (the professor saying, "Right!"). Those who select any of the other responses suffer the punishment of hearing the nutty professor launch off into another one of his tirades.

THE MECHANICS OF INTERACTIVE EDITING

There are two concepts that have special bearing on the mechanics of editing for video discs. These are shot organization and frame flutter. We will begin with shot organization, which was alluded to in Figure 7.1.

Shot Organization

Let's look at three representations of the same video sequence: Figure 8.4 shows an interactive exercise presented in script form. Figure 8.5 shows the same exercise presented as a flowchart, and Figure 8.6 shows it as electronic images on a piece of video tape.

Even though the script (see Figure 8.4) presented the information in the order XQ1, XCF, XFF, XQ2, the editor changed the shot order to XQ1, XFF, XCF, XQ2 (see Figure 8.6). The goal of this kind of editing is to eliminate searching when straight

Figure 8.4: Video Sequence Script

Narrator

XQ1: Identify the following as 1. a debit, or 2. a credit.

XCF: Very good, well done.

XFF: You don't seem to get the idea. Try again.

XQ2: Is cash a debit? Yes or no?

Figure 8.5: Video Sequence Flowchart

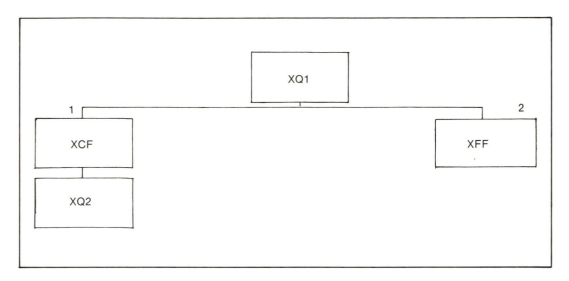

Figure 8.6: Video Tape Placement

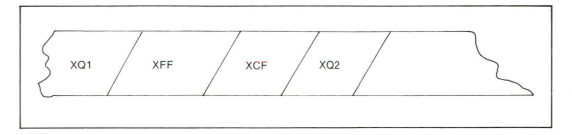

play can be achieved. This is because on many disc players there is a blackout whenever the player searches. By putting XCF next to XQ2, the viewer can go directly from the correct feedback (XCF) into the next question (XQ2).

Figure 8.7 is a flowchart for an exercise using internal branching, so named because an exercise branch is built inside another exercise. The video tape image for the example is shown in Figure 8.8.

The editor tried to eliminate searching by keeping straight play segments contiguous, as in 3CF7, which immediately precedes 3XQ9, and 3FF7.1, 3XQ8, 3XCOM8, which play one into the other.

Generally, shots should be put together in traditional editing order. However, when assembling exercises, the correct feedback should go last, next to the sequence that follows, so that the program plays right into it when a response is made. Looping feedback should be placed nearer the question to reduce search time; questions in internal branching should be grouped together (as 3FF7.1, 3XQ8 and 3XCOM8 are in Figure 8.8).

Figure 8.7: Internal Branching

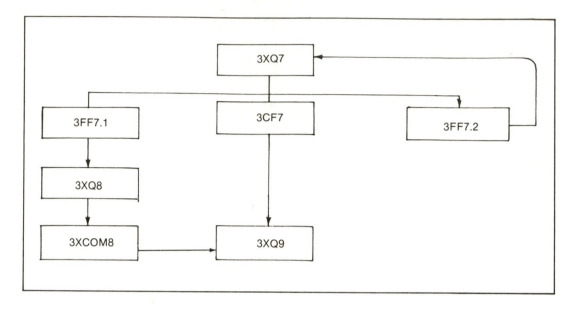

Figure 8.8: Internal Branching: Video Tape Placement

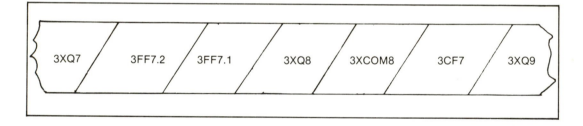

Menu frames and other frames that are accessed frequently should not be placed on the very outer or inner portions of the disc, but more toward the center where accessing is generally more accurate. Of course, menus and other still frames have other idiosyncrasies, chief among which is frame flutter.

Frame Flutter

This is a problem with video discs. As most of us know, the television picture is made up of two fields which interlace. Some editing machines are set to edit on one field, some on the other. It doesn't matter which field has been used when the video program is played without stopping. However, when a program is composed of sequences edited on two machines, or on a machine where the field is switched during the edit, the first frame of some edits is composed of an interlace of fields from two different pictures. When you freeze this kind of frame, the picture appears to flutter between the two images.

Figure 8.9: Eliminating Frame Flutter

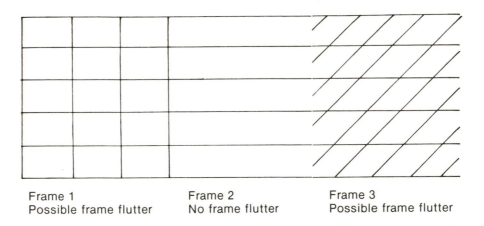

Frame 1
Possible frame flutter

Frame 2
No frame flutter

Frame 3
Possible frame flutter

There are two solutions to frame flutters. First, make sure that you never switch fields when editing. Second, whenever you can, lay down three frames of any image you want to still frame. Then freeze on the second (middle) frame. The result will be that even if there are mixed fields on both sides, the second frame will be solid, as shown in Figure 8.9.

Of course, if you are stepping through a series of stills, you can only lay down one frame of each still; otherwise, you will have to search from frame to frame. In this case correct field dominance is mandatory. It should also be noted that video images always flutter when stopped during any relatively quick motion sequence. This means that you should never still-frame during a motion sequence unless you absolutely have to. If you must, you might consider shooting the sequence on film. Granted, quick motion film sequences show a blurr when still-framed, but a blurred image in motion has long been used as an artistic device. Therefore, it's accepted by viewers, while frame flutter is not.

The editing concepts presented here certainly don't cover all aspects of interactive editing. There are many new concepts yet to be discovered and explored. Rather, I have discussed these concepts since they are the most obvious ways in which interactive video editing differs from standard practice.

9

Consumer Applications

Much of the interactive video development that has taken place to date has been in the area of training and instructional programming. However, in the past few years, consumer applications of interactive video have been generating more and more interest. In this chapter we will examine several applications, including interactive games, point-of-sale systems and video publishing. Specifically, we will look at the design features you must keep in mind when developing these types of programs.

VIDEO PUBLISHING

Interactive video publishing will grow out of a part of the business market that is now alternately called job aids, guidance systems or "cookbooks." These instructional programs have industrial applications, but will also be of interest to consumers as well. Interactive job aids are video information programs that will balance a market now saturated with theatrical films released for home consumption.

To call an interactive job aid a "how-to" program would be a drastic oversimplification. It is a program that presents an overview of a job, tells why it's important to do it right and gives step-by-step instructions on how to do the job *as the viewer is doing it*. Then the program branches to different activities for variations.

When would job aids be used in business? Certainly, when there are not great time pressures involved, when the job is performed infrequently or whenever there is a long series of steps involved in doing something. In short, the difference is between teaching people to *remember* how to do tasks versus having people actually *do* the tasks with a little help.

Interactive job aids have been proposed for such diverse on-the-job applications as routine maintenance, medical diagnosis and product identification. Consumer-oriented subjects suited to interactive job aids include filing your income taxes, home remodeling or repair projects, preparing resumes, cooking a gourmet dish, buying a car. These and many more consumer applications lend themselves to interactive video publishing, and it seems to me that this will eventually turn out to be one of the most popular uses of the medium.

INTERACTIVE VIDEO DISC GAMES

Games are great fun and can also serve as excellent teaching devices. Some games can take a viewer through long processes as complex as an entire economic cycle or military campaign. Others focus on a few special skills such as throwing a ball through a hoop.

The two major kinds of games are action and adventure games. Action games include arcade games and home video games, such as "Pac Man," "Centipede" or "Space Invaders." These games are relatively short and rely on reflex and psychomotor skill. In spite of parental complaints about the environment they create, the fact is that action games greatly enhance hand-to-eye coordination, sense of timing, judgment, etc.

Adventure games, on the other hand, are role-play, decision-making games. They don't take minutes, they take hours or days to play. They rely heavily on imagination. Most home computer games fall into this category, as do a very few home video games like "Adventure" and "Superman."

Let's look at instructional applications of both kinds of games.

Adventure Games

The essence of the adventure game is decision-making. Players often go on a quest, through forests, across deserts, over mountains. They explore dungeons and encounter monsters of various kinds. But what the player is really doing is making decisions: "Do I take this fork in the road or that? Do I enter this dungeon or not?" Often the decisions are informed, for example, "Do I have enough strength, food, supplies and water to continue the quest?" Frequently, there is a tabulation of various materials, weapons, resources.

Basically, however, the purpose of the game is to give the student experience making decisions about whatever principles you—the designer—build into the interactions.

Do you want to teach battle strategy? Set up situations in which battles must be joined according to certain military principles or they will be lost. Do you want to teach conservation? Make supplies and resources limited, requiring careful manage-

ment. Do the same thing with money and you can teach cash flow and money management. Build a series of clues and you can teach reasoning and logic in a "Solve the Mystery Game."

Of course there is one tremendous difference between computer adventure games and interactive video adventure games: In interactive video *you have to show the adventures.* No more rough computer sketches of monsters such as giant ants. If the adventure involves giant ants, you have to deliver something that really looks like a giant ant. If the adventure calls for crossing the desert, you have to find a real desert, or something that looks like one, and cross it.

Once the public starts playing adventure games in which they battle realistic aliens, toss real touchdown passes, plan for the survival of realistic cities, a whole new world will open up. Of course, computers will play a big role in these games. Rays from ray guns will be generated by computers; graphic tablets for computation of supplies or cash flow will be carefully integrated with the video images. The result will be the optimum integration of computers and video, and the ability to teach will be limited only by the designer's ability to make use of the new technology.

Action Games

Action games are the interactive video equivalent of the popular arcade games. While cute cartoon games like "Pac Man," Frogger" and "Donkey Kong" may be developed to their ultimate already, discs could add an amazing air of realism to space-battle games like "Defender," "Space Invaders" and "Missile Command."

Table 9.1 lists several types of games along with the skills they can teach.

Mechanical Problems of Disc Game Editing

The problem for the creators of disc action games is how to provide the action-response connection that currently makes video games so enjoyable. How do you push the button and see the arrows fly, flip the joystick and move your character left or right? The answer is that designers will have to learn to see things differently. Until now, game designers have only looked at the world from one vantage point or camera position—the extreme long shot. "Pac Man" can be thought of as being played with one camera position looking down on the maze. The video disc games of the future will go from long shots of attacking monsters, to close-ups of arrows launching, to extreme close-ups of arrows striking home; from dragon claws, to blurs of action, to tails flying through the air.

Or, to put it another way, video game designers will soon have to learn how to edit shots together (1 + 1 = 3). The ability to move from a wide angle of view to a close-up is not only a unique ability of film-making; it's the way our minds work. Adding this ability to the video game will greatly expand the excitement of the games. It will also solve some of the logistical problems of assembling the images of video disc games.

Table 9.1: Skills Games Can Teach

Games	Skills Developed
Adventure Games	
Mystery	Reasoning Logic
Warfare	Strategy and tactics
Quest	Decision making Math skills
Economic	Cash management Budgeting Principles of economics
Conservation	Decision making
Survival	Planning Decision making Math skills
Simulation	Planning Budgeting Principles of whatever is simulated
Life	Planning Decision making Integrated Practice
Action Games	
Transportation Simulations	Driving skills Bicycle safety skills Rules of the road
Battle	Marksmanship Hand-to-eye coordination
Search	Map skills Hand-to-eye coordination

But, once you make the aesthetic editing decisions for your video game, you still have to deal with the mechanics of the process. If the game is no more than computer images floating over live disc images, there is little problem. But if there is close interaction between the graphics and the disc, or if the game consists largely of branching from one point on the disc to another, you have to deal with the blackout that occurs while the disc is searching. This blackout interrupts the action and disturbs the realism of the game.

New disc players are coming onto the market that will hold an image on the screen during any branching that does not exceed some given number of frames.

Specifically, if you do not branch to a spot more than 200 frames from where you start the branching, you can keep the image on the screen for the split second that is needed for the branching. *There is no blackout.*

A more sophisticated kind of branching occurs when you apply the computer concept of *interleaving* to a disc game. Interleaving works in the following way: disc players have the capability to play every third frame or every fourth frame or, in fact, to skip and play frames at given intervals. By taking advantage of this ability, a disc game can lay down parallel tracks of motion. Branching, then, is no longer a problem of jumping hundreds of frames but becomes a simple matter of moving one or two frames ahead or behind. Figure 9.1 shows what happens in the scene of a man walking down the street. His action going straight ahead (sequence X) is laid down on every third frame; his action if he turns to the right (sequence Y) is laid down on every other third frame; and his action if he turns to the left (sequence Z) is every previous third frame.

Figure 9.1: Interleaving Every Third Frame

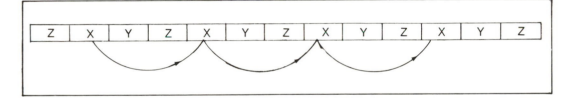

In this example, a decision to turn left or right, as opposed to continuing straight ahead, would only be a matter of jumping ahead one frame (to sequence Y) or back one frame (to sequence Z).

Problems with interleaving as a technique involve the creation of the master tape. Specifically, if you put X sequence on tape machine 1, Y sequence on tape machine 2, and Z sequence on tape machine 3, then you must make a single frame edit from tape machine 1 to record onto tape machine 4 (the record machine). (You can also use an electronic still frame store as a record machine if you have one available.)

Next, you do an edit from tape machine 2 to the record machine, then from tape machine 3 to the record machine, then another series, etc. It's a pretty tedious business. Engineers are working hard to make systems that are better suited to interleaving master material. In the meantime, this technique is still the best system for providing full motion interactive games that has yet been thought of.

We are looking toward full motion games that will allow players to have adventures which they have only been able to view passively before. Figure 9.2 is a flowchart for a sword and sorcery game that will give players great adventures with the highest degree of realism. The game still relies on some computer graphics, but its goal is very much to provide realistic experiences for the player.

Figure 9.2: Sword and Sorcery Game Flowchart

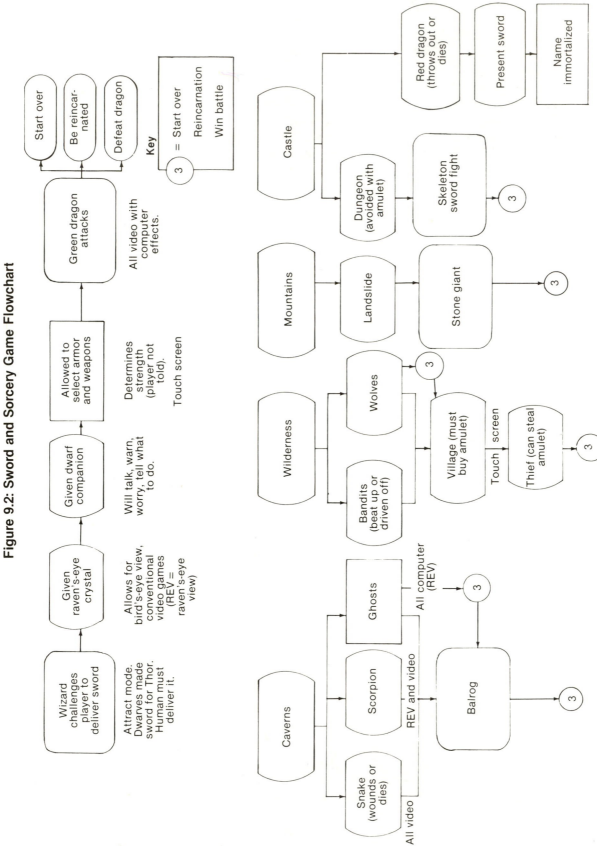

The production costs of such games could be amortized over several applications. For example, a feature film could generate raw material for several game spinoffs for the home and arcade market.

INTERACTIVE POINT-OF-SALE SYSTEMS

In order to develop good interactive point-of-sale material, you have to start with a model for selling. The steps are presented in Figure 9.3. Let's use this model as a mechanism for exploring the sales potential of interactive video point-of-sale terminals. In order to have maximum program flexibility, let's assume that we have one or two disc players hooked up to a microcomputer, with both a typewriter keyboard and a touch screen for customer interface. Assembly of a working unit will not be easy, but recently several companies have begun to specialize in the process (ByVideo, Inc. is one such company).

Every bit as difficult as assembling the hardware configuration is creating the video disc program. Let's look at the sales process and see how it affects the design of our video system.

Figure 9.3: Steps in a Sale

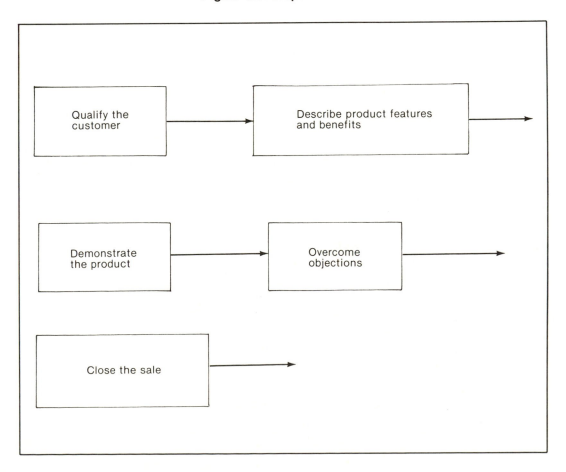

Qualifying the Customer

This normally means making sure the customer is right for the product, but it can also mean finding the right product for the customer. One way to do this is simply to ask the customer what he or she wants. This can be done by presenting a series of menus which guide the customer to the right product.

In banking there are five or six major groups of services; in retailing there are similar product groupings. Begin with menus of product categories and become more and more specific until you've taken the customer right to the perfect product.

What if customers don't know what they want? In this case, a series of specific menus will do little good. Instead, you can ask a series of questions designed to lead the customer to the right product. This, of course, is not an easy design task and will take prodigious amounts of computer memory and creative design skill.

In banking the task is easier than in other sales situations because of all the rules surrounding the various investment instruments. For example, if you can't maintain the required minimum account balance, you simply don't qualify for interest checking. Think of the questions you would have to ask to guide a young man, for example, to the perfect gift to celebrate his first wedding anniversary. Dozens of these kinds of exercises would be needed to provide a qualifying module for any typical consumer video catalog.

Perhaps an easier qualifying method would be to offer customers lavish displays of products grouped in appropriate categories and to let them make their choices from these displays. Such video showcases provide a comfortable way to browse. The customer merely flips through pages of video disc still frames until an interesting product appears.

Whether you present successively refined product lists, questions to guide your customers or lavish product displays, the purpose of this part of the program is to take your customer to a segment about a specific product. Figure 9.4 illustrates the different mechanisms that you can use to lead the customer to a given product segment.

Describing Features and Benefits

In the individual product segments, products are either presented in still frames that are very much like catalog pages or in motion sequences that lead into product stills. We'll discuss product motion sequences in detail, but first let's focus on the still frames.

Still frames should give a full frame representation of the product that gives the clearest view possible. Then they should be followed by product detail frames that offer information about product features and benefits. There should also be a price, delivery, size/selection-type frame.

Figure 9.4: Qualifying the Customer

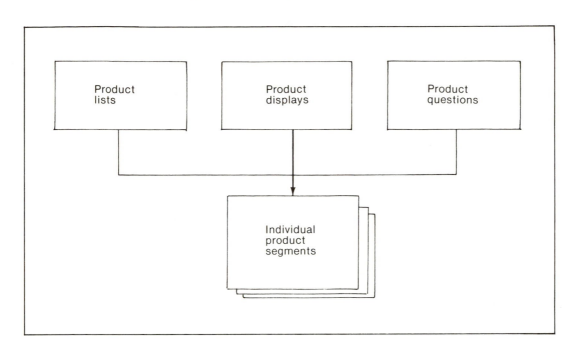

Now let's return to one of the most interesting and provocative questions in all interactive point-of-sale video system design: the use of video motion.

Demonstrate the Product

Various point-of-sale systems call this the "closer look" or the "product commercial." The fact is, while video motion sequences relating to a given product or line of products must have the same high production values as prime time TV commercials, they are to some degree quite unlike commercials. Perhaps the biggest difference is that they are part of an experience that is taking place on the spot. They are trying to get customers to buy something here and now, not next week or the next time they go to the store. Therefore, for writers, producers and directors, point-of-sale video sequences represent a fundamental change from the nature of television commercials.

I prefer interactive video sequences to have very short motion sequences that show the product in action, or quickly set the shopping mood. Very often they can be 3 to 15 seconds in length. Of course, there should also be room for the longer kind of product demo that explains the principles of a technical operation with diagrams or demonstrations. Needless to say, this is the real sales part of the presentation. The product demonstration usually returns to the product still frame to help you close the sale.

Overcoming Objections

One of the first point-of-sale discs produced by Columbia Savings of Colorado offered a section which attempted to overcome objections. The program presented lists of typical questions that customers might have and then offered text responses to the questions. While this approach may offer an economy of disc real estate, a video presentation of a well-prepared spokesperson could do a more credible job of handling objections.

Certainly overcoming objections is not a sequence that needs to be provided with every single product. But there are certain products with which you can anticipate certain concerns. Presenting the buyer with the opportunity to have concerns addressed is a very beneficial use of the video sales system.

Of course, there are opportunities for creative thinking in this particular area. For example, perhaps the option is only offered to the customer who pauses for a certain period of time, who does not buy or who opts to see more products after exploring one very specific product. Or, if the customer hesitates at the product frame, perhaps the objection sequence could begin automatically.

This is a part of the sales process that must be handled carefully, but for which interactive video is surely viable and useful. In fact, in many cases a video point-of-sale system could do this better than most live salespeople can.

Closing the Sale

Even in a sales-informational system, the purpose of the effort is to sell. We've dubbed such uses of interactive video "infomercials," because they attempt to educate and then close the sale. There may eventually be point-of-sale systems that actually do take an order, make a loan, maybe even dispense a product. Those systems must literally ask for the sale (e.g., "Please insert your Visa card now").

However, even point-of-sale systems that do no more than describe the products should refer the customer to a salesperson, an account officer, etc. This need does not grow out of the nature of computers or video or interactivity: it has to do with the nature of selling. Basically, I'm suggesting that you must end a sales effort with a request to buy, or at least a referral to a salesperson. If you don't do that, you've wasted your effort.

Housekeeping

As with instructional interactive video, housekeeping activities must also be offered in point-of-sale programs. The operational procedures have to be explained, and formalities such as welcomes, orientations and farewells must also be attended to. Where the orientations come in your program is somewhat dependent on your products.

Examples of Point-of-Sale Applications

The following section outlines two typical sales/informational systems designed for Bank of America. We'll begin with an outline for a truly interactive point-of-sale disc in Table 9.2. The items with asterisks require full computer capability. The items with two asterisks require the input of a second disc player. Let's look more closely at some of the elements of the program.

Table 9.2: Point-of-Sale Outline

General menu
Motivational welcome
Interactive orientation
Product menus

Checking account products demo
Checking account products qualifier
Check styles catalog

Investment products demo *
Investment product qualifier*
Life game investment simulation * *

Travelers cheques promo
World travel game * *

Video corporate tour
Organizational overview
Glossary of all bank services and
    departments

Wrap-up

*Items require full computer capability.
* *Items require a second disc player.

Investment Products Demo

This is the perfect place for the text overlay feature of computers (text displayed over a video image). The key to investment products is interest rates which, as everyone knows, change all the time. With computer-generated graphics used as text overlay, changing interest rates could be adjusted as often as needed.

Investment Products Qualifier

Checking account products are really very basic and don't require a good deal of complex calculation and cross indexing. Therefore, a simple decision tree might soon lead to the selection of desired check styles. However, investment products are more complex, so a qualifier might require a "fill-in-the-blanks" calculation as follows:

If you put _____ in an IRA for the next _____ years, and
the interest rates were _____, your annual income from the
IRA at retirement would be _____.

The example below shows a similar exercise working backward from goals to contributing elements.

> If you wanted to earn _____ a month at retirement, you could count on _____ from your pension, _____ a month from an IRA (based on _____% of _____ for _____ months). You would still have to supply _____ a month from other sources.

Here are some suggested "other sources":

Again, I urge you not to do this all with computers. Use a warm, friendly human voiceover to explain the formula. That could do a lot to help *sell* the product.

Life Game Investment Simulation

This activity dramatizes some of the consequences of various investment strategies. You begin by making certain investment choices when you start, and make more as the exercise progresses. In the end, the computer tells you what you've managed to put aside and then shows you the kinds of things you will and will not be able to do with your accumulated investment.

In this game many computations are performed by the computer. These computations access various sequences which are selected because they fall within the general range set for each item. Just as life games have been popular for centuries, this interactive video version can be enjoyable, entertaining and very instructive.

An Informational System: The Personnel Information System

A very natural application of point-of-sale systems is the Personnel Education Resource Center which might be located in an employee lounge. A typical system might offer the kinds of activities listed in Table 9.3. The single asterisk items require heavy computer usage. These are items such as qualifiers, plan selectors and calculators that help you determine the effects of an employee savings plan, pension plan or health plans. For example:

> If you become pregnant and experience a normal delivery, you can expect this health plan to pay _____ of the total cost. In _____% of normal deliveries this represents the full cost.

With other plans the coverage would be as follows:

> Plan 1 — Pays _____ or _____%
> Plan 2 — Pays _____ or _____%
> Plan 3 — Pays _____ or _____%
> Plan 4 — Pays _____ or _____%

Table 9.3: Personnel Information System Outline

General menu
Motivational welcome
Interactive orientation
Program menu
Orientation
Submerged action game

Health care overview
Health care plan qualifier*
Health care video workbook*

Vacation plan overview

Pension plan overview
Pension plan video workbook*
Employee investment plan overview
Investment plan selector*
Pension and investment life game**

Retirement plan overview
Retirement plan calculator*
Retirement dream game**

Employee loan overview
Employee loan executive message
Employee loan video workbook*

Miscellaneous benefits overview
Catalog of miscellaneous benefits

Employee gift catalog
Employee gift ordering system**

Submerged menu and benefits

Employee stock options
Employee cash bonuses
Employee stock bonuses

*Items require full computer capability.
**Items require a second disc player.

"What will happen to me if . . ." adventure games can be created to dramatize the benefits of various products or services. In designing such simulation games for point-of-sale systems, it is a good idea to work backward from the outcomes to the causes. For example, in the proposed "life game" "Retirement," outcomes could be shown as five kinds of situations:

1. Total freedom

2. Limited financial freedom

3. Moderate financial limitations

4. Extreme financial limitations

5. Financial hardship

Video segments (perhaps in a documentary style) would be accessed by certain income ranges. The income ranges, on the other hand, would be reached via computer calculations which, in turn, would be fed by viewer input as shown in Figure 9.5.

Figure 9.5: Point-of-Sale Life Game Design

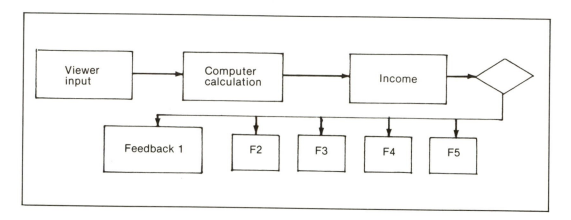

As we have said, action games using interactive video have a fabulous future of their own. But even in a point-of-sale or training environment, computer-video action games might be included in order to interest students in using the system and to give them hands-on experience with the machine. Such games should probably be hidden (submerged) so that only people who know how to reach them via access code can actually get to them.

Other submerged activities might include stock options and cash or stock bonuses that are not available to most employees. Only a benefits expert might know how to reach them.

EVALUATING THE PROGRAMS

Evaluating and testing interactive video games and point-of-sale systems follow the same general principles outlined in Chapter 6. However, there are some important differences, and one or two advantages, especially in the area of video marketing.

First, in video marketing and video games, both a focus group or a pilot test with on-site testers would be practical, and in some cases preferable to long run data collection. On-site testers can see user reaction, hesitancy, interest. A well-constructed pilot conducted by a disciplined market research firm can do a good deal to strengthen the design features of point-of-sale and gaming systems, and to evaluate their effectiveness.

In the area of video marketing, the computer's ability to count becomes a valuable market research tool. How many people looked at products one way as opposed to another, how many asked to be told how to shop, what percentage of all these people were actual buyers—all of this information can be collected and analyzed by the computer, if you design your program accordingly.

SUMMARY

Interest in interactive games and point-of-sale systems appears to be growing in the U.S., Canada and other countries. While established as a viable training and instructional technique, interactive video, like many other technologies, will most likely derive its momentum for growth from success in the consumer market. There is no doubt that the technology will continue to advance, making interactive systems increasingly sophisticated. However, the success of interactive video does not lie in the capabilities of the system, but in its applications. This is the challenge that interactive video designers must meet.

Afterword

In 1979 the world of interactive video was new and bright with promise. Wanting to share in that promise, I stood on the stage of the International Tape/Disk Association (ITA) National Convention—where just the day before the founding of DiscoVision had been announced—and boldly stated that within five years all video instruction would be interactive; within 20 years, all entertainment would be interactive. I should have known better.

My mistake was not that I gambled on interactive video; my mistake was in thinking that training would be the catalyst to bring interactive video to full development. Training is most often a stepchild of industry, since it is at best an indirect source of income to the company.

As I pointed out in Chapter 5, training's direct impact on the bottom line is extremely hard to isolate and measure. As a result, the sudden investment of millions of dollars in new interactive video training hardware is wishful thinking. I'd like to hold to my prediction that eventually all video training will be interactive video, but my timetable clearly needs adjustment.

The expansion of the interactive training system may in fact be tied to the emergence of two more profitable applications of interactive video in the consumer market. The first is point-of-sale video, which is the forerunner of real-image video-shop-at-home systems. People *will* be able to see products on their TV screens and order the products on their home computers. The images will come from banks of video disc players which will search and display product information and promotional messages.

But before video shopping reaches the home, it will appear as stand-alone terminals in showrooms, in catalog departments of general merchandise stores and even in shopping centers. It will offer products for sale through a credit card reader or from a nearby counter clerk. Similar machines will pour out information on local sports events, restaurants, tourist attractions. They will differentiate among kinds of financial services and even demonstrate product uses.

Many of these machines will offer flip-side instructions that tell employees how to sell products and deal with customer inquiries. Thus interactive instruction may, in fact, make it to the marketplace through interactive video shopping. But even as that happens, video discs could have a revolutionary effect on another industry.

The video game business has suffered a massive shake-out. People are bored with the old games and game ideas. Most of the manufacturers of these games would be heading down the irreversible road to bankruptcy if there weren't another path to take. Fortunately, there is. If developed to its full potential, interactive laser disc games can add a level of realism that is astoundingly superior to the limiting graphics of even the *best* computer-based games.

If game manufacturers can avoid the quick fix of gimmicky games that minimize the role of the disc and just aren't fun to play they will find that the future of interactive disc games can be virtually limitless. Beyond the horizon lie the vast realms of mega-simulations, the playroom and other interactive entertainments not yet dreamed of. They are there waiting for someone to discover them. Let's get started!

Glossary*

access: The ability or action of going to or reaching an item on a *video disc* or video tape.

action game: A video game dependent on quick and accurate response by the player. These games are usually of short duration.

adventure game: A computer or video quest/search in which decision making and exploration play a major role. These games are usually of long duration.

branch: The process by which the user is routed to the appropriate part of a program or *excercise* according to his or her responses.

combination treatment: A major video style that can include *dramatic, didactic, documentary* and *graphics treatment* elements in a single program.

computer: See *external computer, on-board computer.*

decision point: The moment in a program where the user must make a choice, such as selecting the correct answer or deciding what action should be taken next.

demonstration: The part of an instructional *exercise* that shows the proper way to perform a task.

didactic treatment: A style of video that features a teacher or spokesperson delivering a message or demonstrating a product or procedure.

disc: See *video disc.*

---

*Words that appear in italics within definitions are defined elsewhere in this glossary.

documentary treatment: A style of video that presents real-life situations, real people doing real things.

dramatic treatment: A style of video that tells a story, acting out examples with fictional characters.

editing: The assembly of shots into a final production.

exercise: An activity in which the user learns a principle, skill or concept by using it repeatedly. Also called practice exercise.

external computer: A computer mounted outside the video disc player; generally more user-friendly and flexible than an *on-board computer.*

feedback: The return of part of the output as input to produce changes that improve performance or provide self-corrective action.

flowchart: A diagram that represents the flow of information in an *interactive video* program.

frame: One complete video image. A frame is composed of two interlaced fields of video information; 30 frames equal one second of motion.

frame flutter: A visual disturbance in a *video disc* image caused when editing creates a *frame* out of fields from two different video images. Can also happen when a motion sequence is interrupted with a freeze frame.

graphics treatment: A style of video in which titles, charts, graphics and schematic diagrams provide all the visuals.

housekeeping: The elements of instruction that deal with operating the exercise or lesson; in *interactive video*, how to use the controls, what to do when you want to take a break, etc.

interactive video: Any video program in which the sequence and selection of messages is determined by the user's response to the material. See also *linear video.*

laser disc: See *video disc.*

linear video: Any video program in which the structure, sequence, pace and selection of material are predetermined and invariable.

manual operation: Action that must be carried out by the operator of the system, apart from standard interaction.

matching exercise: An instructional technique that requires that items in one group be paired up with items in another group.

menu: A list presenting choices to the user at major *decision points.*

on-board computer: A computer that is mounted inside the video disc player.

overlay: The ability of some *external computers* to superimpose text or graphics over a video image, creating a picture that combines both video and graphics.

point-of-sale: The location and time when someone is about to buy. A point-of-sale terminal is available at that place and time to help close the sale; it may in fact take the order. Also called point-of-purchase.

post-production: The work done after *production,* particularly *editing,* special effects creation and character generation. For *interactive video,* this part of the process is very different from *linear video.*

preproduction: The work done to prepare for video or film *production.* Includes casting, location selection, set preparation, prop gathering, shot organization, equipment production, etc. Production is similar for *linear* and *interactive video.*

prerequisite: The skill(s) that one must possess in order to understand a given lesson.

preview: An *interactive video* program unit that offers an overview of the information to be presented in the lesson.

production: The actual shooting of film or video tape sequences for a program.

random access: A method of storing data so that they are directly accessible, independent of their location in storage. Also, the capability of a *video disc* to access specific frames.

sequencing: The order of presentation of information in an instructional program.

simulation: A teaching strategy that presents the user with a true-to-life situation in which the skills to be learned must be used.

spiderweb: A learning *exercise* in which *branching* from different *decision points* may lead the user to some of the same results. The flowchart thus resembles a spiderweb rather than a *tree.*

subroutine: A self-contained unit that is accessed and then returns the user to the main body of the program.

timed video still: A frame of information left on the screen for a specific period of time.

touch screen: A response mode that allows the user to make a choice by touching the proper area on the video screen.

treatment: The form of presenting a subject in film or video. There are five major kinds of treatment: *didactic, documentary, dramatic, graphics* and *combination.* Also, a narrative description of a proposed video program.

tree: A learning *exercise* structured so that one decision leads to additional choices, which in turn *branch* to additional choices. Thus the flowchart resembles a tree. See also *spiderweb.*

unit: All the material related to a given subject in a training lesson.

unit test: A test on all the key information presented in a *unit.*

varied outcomes: The situation that results when the different choices at a *decision point branch* to very different results; used in *interactive video* to show the consequences of different decisions.

video disc: An electronic communications medium that has audio and video images encoded in grooves on a flat surface. The grooves are read by a stylus (capacitance disc) or by an optical laser. Laser discs offer *random access* and are more durable; thus they are better suited than capacitance discs to *interactive video.*

Bibliography

Amada, Genine. "Why Educators Should Take the Interactive Media Plunge."
   E-ITV (June 1983): 45-46, 56.

Bennion, Junius L. "Authoring Interactive Videodisc Courseware." In *Videodisc-Microcomputer Courseware Design,* edited by Michael L. DeBloois.
   Englewood Cliffs, NJ: Educational Technology Publications, 1982.

———. *Authoring Procedures for Interactive Videodisc Procedures - A Manual.*
   Provo, UT: Division of Instructional Research, Development, and Evaluation,
   Brigham Young University, March 1976.

Call, Diane. "Basic Principles for Doing Interactive Discs." *E-ITV* (October 1983):
   98, 103-04.

Conkwright, Tom D. "Computer Screen Resolution: The Overlooked Factor in
   Buying Decisions." *Training* (September 1983): 77-79

"Development & Delivery of Training: Media, Methods and Means." *Training*
   (October 1983): 47-53.

Devlin, Sandra. "Premastering for an Interactive Videodisc." *E-ITV* (November
   1982): 38-39.

Donelson, W. "Spatial Management of Information." SIGGRAPH '78 Proceedings. *Computer Graphics* 12 (August 1978).

Dunton, Mark, and Owen, David. *The Complete Home Video Handbook.* New
   York: Random House, Inc., 1982.

Eisenstein, Sergei. *Film Form.* New York: Harcourt Brace Jovanovich, 1969.

———. *Film Sense.* New York: Harcourt Brace Jovanovich, 1969.

E-ITV 14 (6) (June 1982). Entire issue devoted to interactive video.

Flemming, Malcolm, and Levie, W. Howard. *Instructional Methods Design.* Englewood Cliffs, NJ: Educational Technology Publications, 1978.

Floyd, Steve, and Floyd, Beth. *Handbook of Interactive Video.* White Plains, NY: Knowledge Industry Publications, Inc., 1982.

Gagne, R.M., and Briggs, L.J. *Principles of Instructional Design,* 2nd ed. New York: Holt, Rinehart and Winston, 1979.

Greenblatt, Stanley. *Understanding Computers Through Common Sense.* New York: Simon & Schuster, Inc., 1979.

Harless, Joe. *An Ounce of Analysis (Is Worth a Pound of Objectives).* Atlanta, GA: Guilde V Publications, 1972.

———. *Behavior Analysis and Management.* Chicago, IL: Steipes Publishing Co., 1968.

———. "Performance Problem-Solving Workshop," Available through Harless Performance Guild, Newnan, GA.

Kindleberger, Charles P. "Whither the Interactive Videodisc?" *E-ITV* (October 1982): 60-65.

Langdon, Danny G. *Instructional Designs for Individualized Learning.* Englewood Cliffs, NJ: Educational Technology Publications, 1973.

Lee, Chris. "Adding the New Technology to Your Training Repertoire." *Training* (April 1982): 18-26.

Lee, Robert, and Misiorowski, Robert. *Script Models,* New York: Hastings House, 1978.

Lippman, A. "Movie Maps: An Application of the Optical Videodisc to Computer Graphics." SIGGRAPH '80 Conference Proceedings. *Computer Graphics* 14 (July 1980).

Mager, Robert F. *Preparing Instructional Objectives.* Belmont, CA: Pitman Learning, Inc., 1983.

Merrill, P.F., and Bunderson, C.V. "Preliminary Guidelines for Employing Graphics in Instruction." Paper presented at the Annual Conference of the

National Society for Performance and Instruction, Washington, DC, April 1979.

Mohl, R. "Cognitive Space in a Virtual Environment." Unpublished paper. Cambridge, MA: Massachusetts Institute of Technology, 1979.

Nierenberg, Gerard I. *The Art of Creative Thinking.* New York: Simon & Schuster Inc., 1982.

Panati, Charles. *Breakthroughs.* Boston, MA: Houghton Mifflin Co., 1980.

Pawley, Roger. "It's Becoming an Interactive World." *E-ITV* (June 1983): 80-81.

Performance & Instruction Journal 22 (9) (November 1983). Entire issue devoted to interactive video.

Samuels, Mike, and Samuels, Nancy. *Seeing with the Mind's Eye.* New York: Random House, Inc., 1979.

Sigel, Efrem, et al. *Video Discs: The Technology, the Applications and the Future.* White Plains, NY: Knowledge Industry Publications, Inc., 1980.

The Video Age: Television Technology and Applications in the 1980s. White Plains, NY: Knowledge Industry Publications, Inc., 1982.

Yampolsky, Michael. "The Next Generation of Interactive Videodiscs." *E-ITV* (June 1983): 40-44.

PERIODICALS OF INTEREST

ECTJ Journal. Washington, DC: Association for Educational Communications and Technology (AECT), quarterly.

E-ITV. Danbury, CT: Tepfer Publishing Co., monthly.

Electronic Learning. New York: Scholastic, Inc., 8 issues/year.

Performance & Instruction Journal. Washington, DC: National Society for Performance and Instruction, monthly.

Training and Development Journal. Washington, DC: American Society for Training and Development, monthly.

Training. Minneapolis, MN: Lakewood Publications, Inc., monthly.

Videodisc Design/Production Group News. Lincoln, NE: Nebraska ETV Network/University of Nebraska-Lincoln, monthly.

Videodisc/Videotex. Westport, CT: Meckler Publishing, quarterly.

Videography. New York: United Business Publications, Inc., monthly.

Video Manager. White Plains, NY: Knowledge Industry Publications, Inc., monthly.

Videoplay Report. Washington, DC: Winslow Information, biweekly.

Video Systems. Overland Park, KS: Intertec Publishing Corp., monthly.

Index

ABOUT THE AUTHOR

Nicholas V. Iuppa is an experienced media manager as well as an active writer and producer of interactive video discs. He is currently vice-president of media production at ByVideo, Inc., where he and his staff have designed, written and produced more than a dozen interactive video discs for the ByVideo shopping system. They are currently involved in several design and development projects for interactive informational, game and educational systems, as well as doing ongoing work on video shopping applications.

Previously, as vice-president, director of media services at Bank of America in San Francisco, Mr. Iuppa supervised a large-scale media center where his staff produced over 500 video programs, including the interactive training programs referred to in this book.

Mr. Iuppa's credits as a writer include story development assignments for MGM and Disney Studios and instructional design work at Bank of America and Eastman Kodak's Marketing Education Center. He is also the author of several managerial and technical articles in trade journals, a new book entitled *Management by Guilt* and numerous pieces of popular magazine fiction and poetry.

Mr. Iuppa holds an M.A. from Stanford University and an A.B. from the University of Notre Dame. Both degrees are in communication. He lives with his wife and three children in Belmont, CA.